The Rough Guide to the Energy Crisis

by David Buchan

www.roughguides.com

Credits

The Rough Guide to the Energy Crisis

Editing and layout: Matthew Milton
Proofreading: Neil Foxlee
Production: Rebecca Short

Rough Guides Reference

Director: Andrew Lockett
Editors: Kate Berens, Peter Buckley,
Tom Cabot, Tracy Hopkins,
Matthew Milton, Joe Staines

Publishing Information

This first edition published October 2010 by
Rough Guides Ltd, 80 Strand, London, WC2R 0RL
Email: mail@roughguides.com

Distributed by the Penguin Group:
Penguin Books Ltd, 80 Strand, London, WC2R 0RL
Penguin Group (USA), 375 Hudson Street, NY 10014, USA
Penguin Group (Australia), 250 Camberwell Road, Camberwell, Victoria 3124, Australia
Penguin Group (Canada), 90 Eglinton Avenue East, Suite 700, Toronto, Ontario, Canada M4P 2Y3
Penguin Group (New Zealand), Cnr Rosedale and Airborne Roads, Albany, Auckland, New Zealand

Printed and bound in Singapore by Toppan Security Printing Pte. Ltd.

312 pages

A catalogue record for this book is available from the British Library.

ISBN 13: 978-1-84836-412-7

1 3 5 7 9 8 6 4 2

Contents

Part 3: The players

Part 4: Energy and emergency

Part 5: Energy prospects
What will the future look like?

Resources

Introduction

The energy crunch

Energy is what keeps us – and our way of life – going. It is something we tend to take for granted. As long as there is electricity in the wall sockets and petrol in the pumps, most of us are not worried about how the electricity is generated, or concerned about possible alternative fuels for our cars. This book is a guide to energy: its different forms; the myths that surround it; and its relationship to the economy, environment, prosperity, democracy and conflict. But it is also an explanation of the necessity of moving towards a low-carbon future. The need for this transition is primarily due to climate change, the planetary crisis created mainly by industrialized countries and now exacerbated by developing nations.

There are two billion people in the world who have no connection to a power grid, for whom every night is effectively a blackout. But in the rich industrialized nations, the worst we typically suffer might be occasional power losses when a tree falls on a pylon or a fuse blows. From time to time, petrol or gas becomes expensive. But these interruptions are generally due to geopolitical disputes or accidents affecting power supplies. We expect such incidents to be temporary, energy supplies to resume and life to quickly return to normal. Nothing essential alters.

The onset of climate change, however, dramatically changes the picture for energy. Three crucial fossil fuels – coal, oil and gas – are the mainstays of world energy production, providing around eighty percent of it. But they are also the main culprits of global warming. And, as a global average, our energy use is responsible for two thirds of all man-made greenhouse gases. So the future of the planet's climate and the future of its energy system have now become almost totally wrapped up together. There are very few proposals for mitigating climate change (involving, for instance, less intensive forestry and agricultural practices) that do not relate to energy. And it is, in turn, hard to think of any major energy issue today that does not relate to the climate change problem.

Take nuclear power. The perennial problem of safely disposing of nuclear reactor waste is one that has always made nuclear controversial as an energy supply. But even that problematic aspect of nuclear power has to be set against the undoubted ecological advantages of nuclear, as a low- or zero-carbon energy source: the potential scale of global warming is leading some greens to reconsider it, for purely pragmatic reasons. In other words, impact on the climate has become the universal yardstick against which every energy system has to be measured.

Ticking clocks and rising mercury

The writer and columnist Thomas Friedman has described our age as "the Energy–Climate Era". It is clear to anybody with even a passing interest in world affairs that the clock is now ticking for energy solutions to climate change. For the planet appears to be warming faster than many climate scientists had initially predicted.

A few years ago the United Nation's Intergovernmental Panel on Climate Change was attacked for scaremongering by estimating that the earth was headed for a temperature increase of between 1.8 and 4 degrees Celsius by 2100. Now that estimate looks conservative. The meteorologists working for America's Central Intelligence Agency concur. Today the global average is nearly one degree Celsius above pre-industrial levels, and any increase to more than two degrees could prove disastrous.

"The stone age did not end for lack of stone," pronounced Zaki Yamani, the Saudi Arabian oil minister, in an oft-quoted comment. He was warning countries belonging to the Organisation of Petroleum Exporting Countries (OPEC) that high oil prices might lead to a collapse in demand for oil long before the supply of oil ran out and it increasingly appears that, for the good of the planet, it would probably be a good thing if Yamani's scenario came true. We should drastically cut our current use of fossil fuels – or take radical steps to clean them up – long before we in any way run out of them.

But it won't be easy. Coal, gas and, above all, oil are so fantastically convenient – dense stores of energy that are also reliable and transportable – that we will have great difficulty in shifting ourselves towards a low-carbon economy. Fossil fuels have driven world prosperity forward. Oil, gas and coal have enabled an economic growth that has in turn speeded up the extraction and use of more oil, gas and coal.

This apparently virtuous circle has turned vicious on us. The steady build-up of greenhouse gases, mainly caused by the burning of energy hydrocarbons, could, if unchecked, lead to runaway warming of the world's climate leading to more floods, droughts, disease and a possible permanent loss of between five and twenty percent of world income each year. That economic cost of climate inaction was an estimate by the UK team of economists and climatologists led by Nicholas Stern in 2006. Stern was accused by many at the time of being a climate Cassandra. Today Stern and many others believe the report erred on the side of optimism.

Yes we can

All is not lost, however. It is true that the 2008–09 recession made people think more about their wallet than the planet, and increased the public indebtedness of Western governments. But it also created opportunities for advancing the cause of cleaner energy.

The need for a big public spending stimulus for economies around the world created an exceptional opening for green energy investment. Low-carbon energy investment is not necessarily the quickest-acting or most job-creating catalyst for growth, but it figured large in some government programmes.

The economic crisis shook faith in market mechanisms (including, unfortunately, the emission trading system in which all Europe's big energy users participate). But the crisis also lent political respectability again to regulation and government intervention. The effect of this change in mood was to widen the tools available to energy policy-makers. In this new context, increased use of regulatory instruments, such as energy efficiency standards for appliances or quantity quotas for renewable energy supply, seem more feasible. This book will explore the relationship of energy to the economy, environment, poverty, democracy, war and other geopolitical issues. But it is essentially the story of how developed countries can, and must, lead the way to a low-carbon future.

It is not an impossible journey, although there are powerful myths that make it seem so. Recent events have shown we are not as totally addicted to oil as is thought. The long run-up of the oil price from 2000–08 and its subsequent crash in 2008–09 showed that oil was not the tail wagging the dog of the world economy. The head of BP, one of the leading sellers of gasoline in the US, said in 2009 that his company would probably never sell as much gasoline again in the US as it had at the start of 2008.

Nevertheless, while the necessary scale of change requires governments to give a lead with clear and bold policies, many politicians still seem paralysed by indecision. Why are they so undecided? Partly because energy decisions frequently involve decades-long projects that require long-term policy frameworks that in turn lock countries into certain energy sources and infrastructure. Governments are also hesitant because they are nervous about how their citizens and voters would react to any clear, bold policies. Yet most individuals, at least the better-informed ones in more affluent countries, are probably ready to do their bit for a low-carbon energy economy – paying more for green electricity, driving less – provided their individual gestures add up to a perceptible difference. But that requires governments to set clear and bold policies. The only way out of this impasse – as will be examined at the end of the book – is for governments to come under pressure to act from a public that is ever-better informed about energy and energy choices. So the answer to "who is to lead?" is, ultimately, us.

How this book works

Part One is an overview of the major energy issues facing today's world, laying out the scale of the energy challenge and providing a road map to some of the pitfalls along the way to a low-carbon economy. It also explains how governments around the world are stepping back into energy policy (if they ever really left it) to address two market failures. One is the quasi-failure of the oil market, whose chronic price swings have a huge impact on the world economy. The other is the world's most dramatic market failure: our failure to realise, and rectify, the climate damage wrought by the fossil fuels with which we've built our civilisation. **Part Two** spans the range of energy sources, both traditional and renewable. The focus is on the key question of today for each source: oil (supply and reserves), gas (transportability), coal (pollution), nuclear (safety and expense) and wind and solar power (intermittency). **Part Three** looks at the players – producers, traders and investors – in the energy game, and the big money that goes into, and comes out of, energy. **Part Four** deals with energy security issues, analyzing how energy, particularly oil, has been a corrupting influence inside countries, and a divisive factor between countries to the point of going to war. It warns that more resource-related conflict is likely. **Part Five** looks at how we can try to accommodate all the world's basic energy needs in a habitable climate – using and wasting less energy and generating more of it renewably – and what kind of political push will be needed to get us there. Only through popular mobilization, led by green pressure groups, will democracies vote for change.

Acknowledgements

My most obvious debt of gratitude goes to Rough Guides – to Andrew Lockett for asking me to write, and helping me to conceive, this book, and to Matthew Milton for his skilful editing, improving and delivery of it.

I owe a vote of thanks to Christopher Allsopp and all my colleagues at the Oxford Institute for Energy Studies for imparting their wisdom and understanding of energy issues. Inside the institute, my particular thanks go to Malcolm Keay and Charles Henderson for reading and commenting on the whole draft, and outside the institute to Peter Aptaker for casting his expert eye over the parts on electricity. I hope this book shows that I have also imbibed something from the energy market expertise of my former colleagues at the Argus Media group, and, before that, at the *Financial Times*. I would like to thank the many people I've talked to in writing this book, not all of whom it has been possible to name. Among those named are two – Stephen Tindale and Jeremy Leggett – who deserve another mention for helping me with their insights into environmental politics. Any errors, of commission or omission, are my own.

As always, I have been sustained by the love, tolerance, and actual interest in energy matters, of my wife Lisa, and by the desire that the environmental future of our children, Susannah, Charlie and Julian, should not be compromised.

PART 1

LEAVING THE COMFORT ZONE

Energy and transition

The great escape

Where we are now and where we go next

We hear so much these days about new sources of energy – about wave and tidal power as well as wind, about different forms of solar energy, about making gas from garbage or oil from algae – that it is easy to get the impression that fossil fuels or hydrocarbons are on the way out.

But this is far from the case, as the table below shows. Attempting the great escape from fossil fuels and converting the world to a low-carbon energy diet is an enormous task.

Projected energy demand on present policies

Energy types in million tonnes of oil equivalent	2007	2015	2030	Compound average annual growth 2007–30
Coal	3184	3828	4887	1.9%
Oil	4093	4234	5009	0.9%
Gas	2512	2801	3561	1.5%
Nuclear	709	810	956	1.3%
Hydro	265	317	402	1.8%
*Biomass/waste	1176	1338	1604	1.4%
Other renewables	74	160	370	7.3%
Fossil fuels (coal, oil and gas) as % of total energy	81.5%	80.5%	80.1%	–
* Includes traditional biomass widely used in developing countries; excluding this from the total would make the fossil-fuel share higher still.				

Source: International Energy Agency, *World Economic Outlook 2009*

In rural India, many people are not connected to the electricity grid, and still rely on very basic biofuels such as animal dung to cook with.

Developed countries have the technical and financial means to make the jump, if not the requisite international cooperation. But there is a huge amount of commercial and political inertia in changing their energy systems, which have been tailored to the convenience of extensive use of fossil fuels. Altering our energy-use habits often involves some personal sacrifice. Sacrifice is not something for which many people are ready to vote in the democracies that exist in most developed countries.

For developing countries, the challenge is even bigger. They are being asked to steer a very different course in fuelling their economic development – not along the well-worn fossil-fuel track already taken by industrialized countries before them, but along the steep upward path of energy innovation to a new low-carbon system. The response of developing nations to the developed ones can be loosely summed up as: "After you".

It is impossible to grasp the reluctance of people to move beyond fossil fuels without understanding their extraordinary convenience. The human race has quite literally grown with successive forms of progressively more convenient hydrocarbons. Change in forms of energy has played a part in population growth. Coal for heat, light and power reduced the pressures that earlier dependence on biomass had placed on land for growing wood and feeding draught animals. Oil provided the input for the fertilizers, pesticides and ploughing needed to feed larger populations – quite apart from the synthetic materials clothing them, the fuel for their mobility and the plastics put to every conceivable use. Moreover, electricity – two-thirds of which is generated by hydrocarbons – provides instantaneous energy at the flick of a switch.

It is because fossil fuels are so versatile and such a good store of energy that we seem to be determined to hang on to our hydrocarbon status quo for as long as possible. So it is all too plausible for the International Energy Agency to predict that, without a radical change in present policies, the fossil-fuel share of total energy will still continue to hover around the eighty percent mark by 2030.

At the same time, however, it is not possible to understand the desire of many poor countries to complete their development in the traditional fossil-fuelled way without grasping how the energy status quo is sheer drudgery for many of their citizens. An estimated 1.5bn people have no electricity (see p.101) – two-thirds of them in Asia, the rest in Africa and mostly in the countryside. Most of them do not even have access to the relatively unpolluting products such as kerosene or liquid petroleum gas for cooking.

As much as eighty percent of India's population depends on firewood for cooking. This involves mainly women and children spending several hours a day collecting wood or dung for cooking fires. It is not just back-breaking work and an avoidable waste of their time. It also leads to deforestation and is a waste of resources: dung, for instance, could be used as organic fertilizer.

Worse still, burning wood or dung in indoor stoves in badly ventilated houses kills some 1.3m women and children a year (half of them in India and China), according to the World Health Organization (WHO).

Killers of the world

Cause of deaths	Annual deaths worldwide (millions)
HIV/Aids	2.8
Tuberculosis	1.6
Smoke from biomass	1.3
Malaria	1.2

Source: International Energy Agency, *World Economic Outlook 2006*, based on WHO figures

Ever-increasing energy demand

In the absence of any change in current trends in people's behaviour, or in government policies, the International Energy Agency projects a 40 percent increase in world energy demand over the 2007–30 period. Although lower than was projected before the 2008–09 recession, this is a massive rise.

In terms of fuel, the biggest increase will be in coal, and in terms of national usage, the biggest increase will be in China and India, which together will account for 54 percent of the total jump in energy demand over 2007–30. These two trends are tied together. On current form, China is likely to double its energy use over the next 22 years, and the bulk of this increase will be in coal. Overall, most – around ninety percent – of the future growth in energy demand will come from developing countries. This is because they are undergoing an increase in the three big factors that determine energy demand: population, wealth and mobility.

Population

One of the most important and obvious determinants of energy demand is population. According to the United Nations Population Division, the world's population will increase from around 6.8bn today to 9.2bn by 2050; at this point it may stabilize, but this will be too late to even come close to stemming climate change in and of itself.

It is already a stretch to imagine how we are going to carry out a clean-energy revolution for *today's* level of population – of whom 1.5bn have yet to be connected to an electricity grid – let alone contemplate how to extend this revolution to 2.4bn more energy consumers over the next forty years. The expected growth in world population – a further third by 2050 – means that we are trying to hit a moving target of steadily rising energy-demand.

The link between energy and population is complex. It works both ways – that is to say, energy has also helped increase population growth in the past. Of course, there were many non-energy factors, such as advances in medicine, involved in the population explosion that accompanied Western Europe's industrial revolution. But it is the case that when the human race used only biomass – from the beginning of time to around 1850 – the population multiplied only very slowly. Extrapolating the population growth rate of the "biomass era" for conjectural purposes would give us a gobal populace of just over 1bn today, compared to the 6.8bn of us currently alive. Widespread use of coal from the mid-nineteenth century onwards speeded up population growth considerably. On one estimate, the "coal population" level – the number of people the world could sustain if coal were the only energy source – would be around 2.3bn today, a third of the current population. This hypothetical coal population is harder to estimate than the biomass population because quite quickly – by the early twentieth century – coal ceded its energy primacy to oil.

But just as movement *up* the hydrocarbon ladder – from wood to coal, then from coal to oil and gas – helped increase the world's population in the past, so movement *down* the hydrocarbon ladder – by using less fossil fuel and more wood/biomass – might reduce population because of a return to less energy-rich fuels. It is entirely possible that climate change itself will have a dramatic demographic impact.

For catastrophic climate change would spell an extreme downward pressure on population, due to floods and droughts. James Lovelock, the environmentalist and "Gaia theorist", forecasts that climate change will be catastrophic and that it may well kill several billion people.

Population control is always a controversial issue, and its emergence in the energy/climate debate is no exception. For instance, an organisation called the Optimum Population Trust (OPT) has proposed a scheme called PopOffsets, whereby people can offset their carbon footprint (from, say, air travel) by making a financial contribution to "funding the un-met need for family planning".

OPT's rationale is that the fewer emitters there are in the world, the less emissions there will be. But the scheme has been criticised for a neo-colonial approach: its long-term solution effectively allows the rich to maintain their lifestyle by restricting the number of the poor in developing countries, who have the greatest "un-met need for family planning", but who often contribute least in per capita terms to emissions.

A more common-sense approach is for governments to make responsible birth control – meaning citizens' access to informed but voluntary means of contraception – part of their climate change (and therefore, indirectly, energy-use) strategy. This is all the more necessary because the basic energy needs of today's 6.8bn population have to be taken as given in any future climate regime. Each of us deserves the modicum of energy needed to sustain a reasonable life.

Wealth

Over the past two hundred years, the world's population has grown six times, but energy consumption per person has risen by a factor of eight. This discrepancy is due to rising incomes: the more money we have, the more energy we tend to use. So it would appear that the amount of energy each person uses, which is determined largely by their income, is a lot more important than the number of people on the planet requiring energy. Generally, there is a stabilizing point: once a certain level of relative wealth has been attained by people within a given country, they will eventually spend a smaller relative share of their money on energy consumption, so ultimately this rise in their energy-use tails off. But this tailing-off only occurs at a very high level of both energy use and income, well above the level China and India are expected to reach in the near future.

The middle classes of China and India will go on aspiring to US levels of consumption and, until they reach these levels, they will continue to use more energy. In China's cities, rising incomes have raised energy demand, as newly well-off city dwellers have treated themselves to cars and domestic appliances. In contrast to city residents in OECD countries, whose energy consumption is less than those living in the countryside, China's urbanites consume more energy than rural Chinese, for the simple reason that they are much richer.

The result has been to blunt the effect on energy demand of China's population-control policy. Through the one-child policy (which has now been modified to allow parents to pay a fee for having a second or third child), it had succeeded in slowing the rate of population growth, from an average annual rate of 1.5 percent between 1980 and 1990, to 0.9 percent between 1990 and 2005, and a projected 0.4 percent average annual rate from 2005 to 2030. So even in the one country in the world with a policy of regulating or taxing population growth, the energy impact of that policy has been undermined by income growth.

Keeping people in poverty is clearly not the solution to the fossil-fuel crisis for many reasons. For one thing, energy is essential to achieve a number of development goals, such as providing people with water, food and health care. For another, the renewable and low-carbon energy sources needed to replace hydrocarbons are usually much more expensive than the hydrocarbons – poorer nations will need subsidies and support from rich nations in order to afford them.

And adapting to a warmer global climate will cost money. Indeed, the argument of climate-change sceptics such as Bjørn Lomborg is that, if man-made global warming is actually occurring, money is better spent on adaptation than mitigation. In other words, countries would be better advised not to throw their money at a climatic trend they cannot prevent, but to focus on straight economic development and use their resulting extra wealth to build defences against rising ocean levels – such as higher sea walls around their port cities.

However, the probable scale of climate change's effects – as examined later in this section – make these economic arguments seem more than a little short-termist.

Mobility

Mobility is perhaps *the* defining feature of our globalized economy and way of life. Raw materials and goods are constantly being shipped around the world. The Chinese economy depends on imports of raw material and uses them to churn out manufactured products that are exported throughout the globe. All this transport and transformation of materials and goods increases energy use. But mobility is of course by no means confined to produce. Economic migrants move to other countries to better themselves, become richer and, as a consequence, use more energy.

Then there is the personal mobility revolution provided by the car. As with everything else, car-use is proceeding apace in China. The number of vehicles there rose from 5.5m in 1990 to 37m in 2006. Nearly two-thirds of these were private cars, and until the world-wide downturn of 2008–09 this number was accelerating. The annual increase in new-car sales in China between 2000 and 2006 was 37 percent, and China surpassed

The world's least expensive automobile, the Tata Nano, is often referred to as the $1000 car (although its price tag is actually about the equivalent of US$2500).

Germany (in 2004) and Japan (in 2006). Finally, in 2008, China overtook the US to become world leader, with 9.3m sales of new vehicles (cars and trucks), compared with 8.7m new vehicle sales in the US. China increased its lead in 2009 with an amazing 46 percent rise in new vehicle sales, reaching 13.6m, as compared with 10.4m for the US.

On current trends, there could be 250m vehicles – the same as today's entire US car fleet – on China's roads by 2030. This would be a huge total addition to world car emissions, even though China has introduced fuel efficiency norms that will reduce emissions from individual cars. As with everything in China, numbers tend to overwhelm unit efficiency. India already has more vehicles on its roads than China. It doesn't have the same rate of growth, but to maximize sales to the less well-off, India's Tata group has introduced the so-called $1000 car.

It would be very hard to put the brakes on the mobility revolution and to dampen the demand for personal freedom that having one's own car seems to offer – traffic congestion aside. Developing countries will go through the same transport revolution that developed countries did. The car can't be uninvented; it can only be reinvented.

And, of course, the travel revolution has now long since taken to the air in developed countries, with budget airlines providing cheap flights: demand for similar services from the middle-classes of the developing world is only to be expected. In theory, modern technology such as teleconferencing, a useful electronic subsitute, ought to replace some of the need for travel. But the evidence is that, in practice, one is rarely a substitute for the other. Indeed, it may be that they are not substitutes, but go together. For teleconferencing or videoconferencing appears to work best between people who have already met each other in the flesh.

So, unfortunately, there may be some truth in the claim of the British Airways' Face-to-Face campaign, launched to sustain business traffic during the recession, that "tangible, human connections are a crucial driver of business growth".

Supply and sustainability: what we use and how we use it

Now we know what factors are contributing to the world's spiralling energy demand, it's worth examining what energy we currently use, in order to assess the problems in supplying it sustainably. Because this book's main focus is on *sustainable* supply, the chart below shows how the various sectors rate in terms of CO_2 emissions.

Main sectors responsible for energy-related CO_2 emissions in 2007 (in percentage of total energy-related emissions)

Electricity generation	41%
Transport	22.9%
Industry	16.5%
Residential	6.5%
Other energy sectors (i.e. oil refining)	4.9%
Services	3%
Agriculture	1.5%

Source: International Energy Agency, *World Energy Outlook 2009*

Coal's comeback: the power problem

Let's start with coal. As the table above shows, generating electric power creates over forty percent of all energy-related emissions of CO_2. The main reason is that over forty percent of all electricity is produced by burning coal, the dirtiest of fossil fuels. And, on present trends, coal's share of global electricity generation is likely to continue to creep up. It is the one fossil fuel that the two most populous countries – China and India – have in abundance. And they intend to make the most of it. The IEA forecasts that, between 2007 and 2030, coal-fired generation will grow by 2.5 times in China and by 3.5 times in India. The results, for all of us, could be dire.

Consuming our solar capital

For the longer term, there was always going to be a problem of sustainability with fossil fuels, even without the global warming problem. Except for geothermal heat and the decaying radioactivity of minerals like uranium from the interior of the earth, all energy sources come from the sun (or moon, in the case of tidal power). But some are constantly replenished by solar radiation, such as wind, wave and of course solar power itself, while others coal, oil, gas are stores of solar energy that have fossilized over millions of years. Even if the necessary heat and pressure for the creation of such fossil fuels could be replicated – which they can't be – it would take centuries to replenish these stores of solar energy.

So in using fossil fuels, we are, as the science writer Vaclav Smil puts it, drawing down on our solar *capital*, a never-to-be-repeated inheritance that cannot last forever. What we should be doing, regardless of climate change, is living off our solar *income*, the winds, waves and solar photo voltaic cells that are replenished by solar radiation, like pay cheques replenishing a bank account. This solar income does not arrive as regularly as we would wish, but the intermittency of wind, wave and even solar power is more than compensated for by the fact they do not contribute to global warming as burning up the stocks of our solar capital does.

The burning of coal, in fact, created an environmental movement well before climate change was considered a risk. Action had to be taken to curb coal-use to mitigate smog in cities such as London, and also to deal with the problem of emissions from the sulphur in coal in areas such as America's upper-Midwest and in Germany and central Europe. Falling to earth as acid rain (rain, or any other form of precipitation made especially acidic by sulphur dioxide), these emissions had the effect of stripping trees of their leaves. In the 1980s, when the share of coal in power generation declined and that of nuclear, natural gas and renewables rose, the world's energy sector actually started to "decarbonize" a little. So it is possible: there has been a modicum of decarbonization in the past.

But this positive trend began to slow and reverse in the 1990s, and the energy sector has, thanks to the big burners of coal – which include the US, Russia, Japan as well as China and India – started to re-carbonize. This is why a great deal of faith is being placed in a generic technology known as carbon capture and storage (CCS) as a way of cleaning up coal. As its name suggests, CCS (see p.81 for details) involves taking carbon dioxide out of coal before or after it is burned, and storing it underground. There is, however, a risk that some of this faith may be wishful thinking:

CCS is the one technology that could be termed green, but which does not require us to move out of our "energy comfort zone" of fossil fuels.

Oil and mobility

Transport accounts for 23 percent of all emissions, and is the fastest growing source of them. The transport sector is also uniquely hard to decarbonize, because it is comprised of hundreds of millions of internal combustion engines, and is nearly totally (95 percent) dependent on one fuel source – oil. The one success story in weaning transport off hydrocarbons has been the electrification of trains, which can be run off centralized low-carbon power sources like nuclear reactors.

Transport and oil: current policies

Transport fuel	% of total in 2006	% of total in 2030	Average annual % growth 2006–30
Oil	94.5	91.9	1.4
Biofuels	1.1	3.7	6.8
Other fuels	4.4	4.3	1.4

Source: International Energy Agency, *World Energy Outlook 2008*

Of all fossil fuels, oil is the most convenient. And no one should completely bemoan its use. The commercial development of oil in the mid-nineteenth century saved from extinction the sperm whale, which up to that point had been hunted for the oily substance used for lighting in its dome-like head. A small but very useful portion of total oil consumption goes into plastics, synthetics, medicines and foods. However, we are currently locked into using it for the majority of our transport, and we need to reduce our dependency by using less of it (in more efficient combustion engines) while at the same time developing alternatives (such as electric cars).

There are some geological constraints to fossil fuels – more so for oil and gas than for coal, which is spread fairly abundantly across the world. There is a long-running controversy about peak oil (see p.51), by which is meant the anticipated peak in oil reserves and production. Purely in terms of supply and demand, it is unlikely that oil production will peak in the next twenty years or so, if only because any seriously high price for oil will accelerate production of the huge amounts of unconventional oil that lie in Canada and Venezuela. (This, of course, assumes that the governments

Rising resource nationalism

The very value of fossil fuels has always fostered possessiveness among the haves and envy among the have-nots. It is somehow hard to imagine one country invading another for its wind-power farms, as Saddam Hussein's Iraq invaded Kuwait for its oil. Indirectly, the possession of oil may make a state somewhat less politically stable internally. For there is a strong correlation, inside petro-states, between oil and autocracy and corruption. This is the so-called oil curse (see p.100 in Part Four). Coming straight into government coffers (as it usually does), oil revenue lays politicians and bureaucrats open to the temptation of corruption. As a highly valuable revenue source, it can remove the need for governments to tax their people (which is nice), but also the need for governments to justify taxation through democratic representation (which is not so nice).

The history of the international oil industry is one of early dominance by the Western oil companies until the early 1970s, which turned out to be a stormy decade when countries belonging to the Organization of Petroleum Exporting Countries (OPEC) flexed their muscles, raised prices and nationalized some of the Western companies' assets on their soil (see p.154). Today, dominance lies with national oil companies. These state NOCs – as they are always called in the oil world – hold most of the world's reserves and account for a share of world oil production that is expected to rise from 57 percent in 2007 to 62 percent by 2030, according to the International Energy Agency (IEA). This has an implication for supply. Increasing amounts of oil and gas acreage are off-limits to the Western oil majors.

The priorities of the state companies are frequently very different to the majors. A country's national company is almost always in less of a hurry to develop its oil and gas fields than an outside company would be, and may well pursue a more cautious policy on the rate of extraction from oil or gas fields. They may also, perhaps surprisingly, have fewer means to invest in extraction than their private-company counterparts. For they are liable to be told by their government shareholder to subsidize fuel prices out of their profits, or divert oil profits to other parts of the economy. As for OPEC, cutting production to raise prices is a regular and open part of the cartel's policy and practice.

and electorates of nations holding these reserves will be happy, from an environmental viewpoint, to exploit them.)

Nonetheless, there is every reason to believe that the growth in oil output will hit a plateau over the next two decades. In the Middle East, oil is relatively easy to extract, though it is often off-limits to foreign companies and rigorously controlled by governments which are usually members of the Organization of Petroleum Exporting Countries (OPEC). Outside the Middle East, oil is getting harder to find and to exploit.

Goodbye to easy oil

Reported global oil reserves have risen fairly steadily over the years. But more recently this has been due to revisions made to estimates of reserves already in production (or undergoing appraisal), rather than actual new discoveries. Estimating reserves is almost more of an art than a science. For there is no standardized international approach to the categorization of reserves according to the certainty of their existence and the profitability with which they can be extracted. Yet profitability and commercial potential play a part in estimating reserves. So as the oil price goes up, so do reserves. Enhanced oil recovery techniques also enhance the commerciality of reserves.

The increase in reported reserves should not be read as any indication that oil is getting easier to find. Far from it. The number of new discoveries is falling off in number and size. No new "elephant fields" have been discovered in recent years.

Rates of oil recovery (the percentage of a total reserve that can be extracted) and of oilfield decline (the pace of falling production) have both increased. This is no surprise. The more you can suck out of an oilfield, the faster it will normally decline. Moreover, production from smaller oilfields tends to decline faster than from big ones.

Saudi Arabia's Ghawar field, discovered in 1948, is still by far the biggest field ever discovered, and in 2007 was producing 5.1m barrels a day – or seven percent of the world's conventional oil. In its 2008 annual survey, the IEA surveyed 580 of the world's largest post-peak fields (i.e. fields whose production had started their decline).

It found an average annual decline rate of 5.1 percent, ranging from 3.4 percent for super-giant fields to 10.4 percent a decline for the merely large fields. The result of the oil industry's need to run faster just to stay still is that decline rates have become more important in determining its investment needs than ups and downs in oil demand.

The transition

and the tools to make it happen

It's clear that power generation and transportation are indispensable, and that fossil fuels cannot sustain them indefinitely – even without factoring in climate change. It should be equally clear, therefore, that a transition to secure and sustainable energy is necessary, and probably inevitable.

Will this transition be smooth? There is no guarantee. The transition could be rough if countries were to compete for ever-higher-priced reserves, while simultaneously making inadequate preparation for alternative energies. The oil companies of Asia's two giants, China and India, have certainly stepped up their search for hydrocarbon concessions around the world. Europe and the US are better placed. They have oil majors of their own, pipeline connections to well-resourced neighbours (Russia for Europe, Canada for the US), and sufficient money and technology to develop alternative energies. In terms of economic power, whether countries manage their energy transition well or badly could determine their relative ranking in the new world order.

But managing the transition is far more than a scramble for, and a hoarding of, remaining oil and gas resources. Indeed, too much backward-looking focus on fossil fuels could distract attention from proper preparation for the future. This will involve governments using various tactics and techniques to change the behaviour of both energy consumers and producers. Information and exhortation can lead to people being persuaded to cut back their energy use or to switch to cleaner energy supplies.

However, the level of voluntary response from private individuals is unlikely to be very high without some degree of government intervention. Examples of such actions are the statutory requirements that new electric appliances must carry information about their relative energy efficiency, to guide the energy-conscious buyer. But it is left to the buyer to decide whether to act on that information.

Very exceptionally, a private energy-saving initiative can influence governments. The leading politicians of Britain's three main political parties have joined many individuals and organizations in the 10:10 campaign to pledge to reduce emissions by ten percent in 2010, though it remains to be seen with what result.

Financial incentives and penalties

These are a crucial bridge to a new energy world because, if applied correctly, they give renewable energy sources some of the advantages of cost that fossil fuels currently have over them. The most obvious way of doing this is to make fossil-fuel users pay for the environmental damage they cause. The successful implementation of low-carbon energy sources, such as renewables and nuclear power, or decarbonizing techniques like CCS will depend on subsidies to make them competitive with fossil fuels, or fossil fuels being taxed to reflect the cost of the carbon they pump into the atmosphere. Or some combination of both – subsidies for alternative energy and more tax on fossil fuels.

Subsidies

In addressing the world's energy problems, it is of course important not to subsidize the wrong thing. The European Union has become a leader in renewable energy, largely as a result of the fixed prices or feed-in tariffs which 18 of its 27 member states have guaranteed their wind-and-solar-power producers. They include Europe's three most successful renewable energy producers – Germany, Denmark and Spain. Many governments around the world, particularly in oil-producing countries, subsidize oil products by keeping the prices of these products very low. It is understandable that governments should want to try in this way to help their poor, or boost local jobs and industry. But the effect of these huge oil-price subsidies, which might amount to as much as $300bn a year when the market price for oil is very high, is almost wholly counter-productive.

Keeping oil-product prices artificially low can encourage a blasé or wasteful attitude to petrol on a consumer level. It also discourages the refining of oil. This is why two major oil-producers, Iran and Nigeria, often have to *import* petrol: they have the oil, but not the incentives to refine it into petrol. Instead of subsidizing oil, which benefits rich petrol consumers as much as poor consumers, these countries would do better to maintain market pricing for energy but to provide financial aid targeted at the poor.

Taxes

Unlike the countries mentioned above, most nations do not subsidize petroleum products, but tax them instead. This tax is at a high rate in Europe and a low rate in the US – which entirely explains the better fuel efficiency, emission reduction and even profit record of Europe's car industry compared to Detroit's.

But Europe has not succeeded in extending taxation to other forms and uses of energy. In the early 1990s, the European Commission proposed an EU tax on all uses of energy – not just energy used as fuel but also energy used in industrial processes and power generation – in a way that would penalize carbon-intensive energy sources. In other words, a carbon tax. But it was blocked by EU governments, even though some Nordic members of the EU have gone on to adopt a carbon tax at a national level.

Cap-and-trade schemes

This is the alternative favoured by most politicians in Europe and a few in the US, chiefly because it is a less overt, and therefore more politically palatable, form of energy taxation. The EU has had its Emission Trading System (ETS) since 2005. The idea of a cap-and-trade system is that the government sets a cap or ceiling on the overall volume of GHG emissions it will permit, and within this overall cap, the government allocates (either administratively or by auction) emission permits to industrial sectors or companies, which are then free to trade these permits on the ETS market if they want to. This gives flexibility for companies to expand (which means they must buy extra permits), diversify or reduce (which means they have excess permits to sell) their carbon emissions.

In essence, an emission-trading scheme establishes a fixed emission-reduction volume and lets the market determine its cost, while a carbon tax fixes the cost of carbon reduction and lets the market determine its volume. Carbon-trading schemes are probably easier to negotiate than carbon taxes, and applied internationally they can give tradeable credits and incentives to find sustainable energy resources to developing countries that would never accept a carbon tax on themselves. But during the 2008–09 recession, reduced demand for emission permits on the ETS lowered the permit price to a level at which it was probably useless as a penalty on fossil fuels (and as an implicit subsidy for renewables). In these albeit exceptional circumstances, many hankered after the carbon price certainty that a tax would have given.

Regulation

Progress towards greener energy sources can also be stimulated by regulation. This could, for instance, dictate that a product should meet a minimum standard of energy efficiency, or that a certain percentage of energy that a company produces should be renewable. Increasingly, developed countries require minimum efficiency standards for a wide range of products, ranging from gas boilers to electric appliances, washing machines, consumer electronics, light bulbs and so on. Even though many of these products are now made in developing countries, the most important market for them is in developed countries, which can use their buying power to impose their standards. In the UK, for instance, the Energy Saving Trust has developed an "Energy Saving Recommeded" certification that appears on all new white goods. The EU has its own "European Union Energy Label", which all new washing machines must carry, and a small number of high-quality machines can even carry the "European Eco-label".

Regulation can achieve spectacular results. The best example of how effective they can be is that of the US Corporate Average Fuel Economy (Café) standards, first introduced in 1975. This enormously improved the fuel efficiency for all US vehicles – from an average of 12.9 miles per gallon in 1974 to an average of 25mpg by 1981. But this rate of progress was never sustained because the Café standards were never seriously tightened. After a long stagnation in US conservation efforts, president Barack Obama proposed a renewed tightening of Café standards in May 2009, aiming to raise the minimum fuel efficiency standard by 2016 to an average of 35mpg for all vehicles. The president had been strongly urged to make this move after his administration's March 2009 bailout of the bankrupt US car companies.

Although regulation has so far tended to be mostly focussed on efficiency standards and energy conservation, governments are increasingly intervening to influence the mix of different energy sources and set targets for low-carbon forms of energy. The EU has adopted a target for a minimum twenty percent average renewable share in all EU energy by 2020, with national targets for its 27 member states. Within this, there is a sub-target for biofuels. Many individual US states have similar renewable

Old "clunkers", marked-up for exchange as part of the US's big automobile upgrade scheme.

energy targets, and the US may adopt a federal goal for the whole country. Expect more of this direct intervention if persuasion or financial incentives fail to deliver sufficient low carbon energy.

The great green energy stimulus
How real was it?

At the onset of the 2008–09 recession, many governments around the world boasted that they had included a large "green stimulus" in their economic recovery programmes, as a boost to the environment and to clean energy, as well as to their economies.

According to a November 2009 report by the HSBC bank, pledges for "green stimulus spending" spanning renewables, carbon capture, energy efficiency, low-emission cars, rail networks and electricity grids amounted to $218bn in China, $118bn in the US and $60bn in South Korea, with much smaller amounts announced by European governments ($13.8bn by Germany, $6.1bn by France and $5.2bn by the UK). However, it has to be admitted that turning announcements and pledges of green stimulus spending into actual disbursement of money has proved slow.

To put this in context, all of the bigger green funds were for disbursement over several years, with the exception of China, which rapidly spent money over the course of 2009–10. Moreover, while these green stimuli covered the sorts of investments you might expect – renewables, other low-carbon energy (clean coal and nuclear power), energy efficiency measures and low-emission vehicles – they were defined widely enough to include money for rail expansion and water/waste treatment as well. Nor was all this money really new and additional to governments' previous plans. Much of it is clearly funding that was already planned but which was brought forward in time, as in the case of the $99bn that China committed to spend on rail expansion in 2009–10. Still, the idea of spending public money on progress to a low-carbon economy – something that the world's governments know they will have to spend money on sooner or later anyway – was clearly a good idea.

It was also a golden opportunity to speed up the usual snail's pace of capital stock turnover in energy. This slow turnover is not surprising. Energy projects involve much preparation, negotiation, planning permission and large amounts of money – so operators want to run these plants for as long as possible to get a payback for their money and effort. Change can also be fantastically slow in energy-using capital stock, such as housing. In the absence of the destruction of war or the rapid growth of China's big cities, the typical turnover of housing stock can be anywhere from 40 to 400 years.

The International Energy Agency has estimated the typical life-cycle of industrial buildings as being between 10 and over 150 years, large hydropower dams at 60–120 years, coal-fired plants at 40–60 years, nuclear power reactors at over 40 years and getting longer, power grids and gas pipelines around 40 years, and cars and household appliances at 10–20 years. Not everything lasts forever: the old-fashioned and energy inefficient incandescent light bulb is short-lived. Though, thankfully, many countries are already phasing these light bulbs out in favour of more efficient and longer-lasting fluorescent light bulbs.

But the only piece of energy equipment whose turnover appears to have been directly accelerated by the stimulus was the car – and by measures designed more to help the car industry than the environment. In 2009 many governments introduced programmes dubbed as "cash for clunkers" in the US and generally as "car scrappage" schemes elsewhere. These gave people a cash payment to scrap their older cars provided they bought a new and usually more energy-efficient car.

The programmes had flaws. Only some governments specifically required that the new cars bought should be more fuel-efficient than the old cars turned in. France, Italy and Spain imposed this condition. The US, Germany and the UK did not, although most new cars of a comparable size will use less fuel than an old one. In the US the average mileage of clunkers scrapped was 15.8 miles per gallon, and of the new ones bought 24.9mpg. A further flaw of Germany's big scrappage scheme was the lack of any enforced destruction of old models. So some German clunkers were reported to have been illegally sold on to African and Central Asian markets, thereby lessening the scheme's benefit to global emissions despite its undoubted benefit to the German car industry.

All stimuli create a slowdown when they cease. For instance, the US programme ran during July and August 2009 and gave people $3bn towards the cost of buying nearly 700,000 new cars during that period. But US cars sales fell sharply in September (though they recovered later). It will generally be very hard for deeply-indebted governments not to stop in the recovery what they started in the recession. Most countries will have to squeeze public spending very hard, and it is impossible to imagine energy programmes escaping the axe altogether. Yet it is just such start-stop approaches that shake investors' confidence in governments' ability to set a long-term climate-change framework and stick to it.

Energy: who controls it?

Too important to be left to the market

Energy is special. It's what keeps us and our way of life going. Because it is an essential commodity, governments usually take some responsibility for maintaining it. Because it is an essential commodity, governments generally like to tax it and occasionally to subsidize it.

In recent years there has been a widespread move to treat energy as an ordinary commodity like any other, so as to introduce more competition in the producing and trading of it. It is, after all, a commodity in the sense of being standardized, interchangeable collections of molecules and electrons.

The changing attitude of governments and businesses towards energy has a historical precedent: salt. While salt was still prized as the sole preservative of food, some governments as in ancient China or in British India decreed its production to be a state monopoly or imposed a special tax on it. But today, salt is a banal commodity, produced and traded everywhere.

Lowering the barriers to cross-border energy investment and trade meant national energy companies surrendering their national monopolies in their national markets. This barrier-lowering process – going under the name of liberalization or deregulation and often involving privatization – was also promoted to developing countries, in the name of efficiency, by international institutions such as the World Bank.

A brief history of deregulation

In the past quarter-century, energy has become a lucrative global business on the back of liberalized markets. Oil, the source of the majority of our energy consumption by some length, has always been a worldwide commodity because of its potency, versatility and transportability. Prior to the 1980s its price was directly set by a cabal that initially consisted of a few Western companies and then a few oil-producing governments. Nowadays – despite the undoubted indirect effect of OPEC's production policies (see p.28) – oil is essentially priced according to whatever the two huge oil futures markets of the New York Mercantile Exchange and London's International Commodities Exchange determine the value of oil should be. There are no unified world markets, in the same way, for other energy sources. But that has not stopped gas and coal being shipped around the globe and, alongside electricity, widely traded in the big regional markets of North America, Europe and Asia.

Energy liberalization began in the late 1970s when the US started to try to free up its energy markets. The Carter administration removed price controls on natural gas in 1978, and a year later – in response to the Iranian revolution and surging international oil prices – removed domestic oil price controls. (Although it is worth remembering that at the same time, the Carter administration did foreshadow the Obama administration's interventionism by creating a government corporation to make synfuels – fossil-based alternatives to petrol.)

However, federal attempts to deregulate the US electricity industry have always run into opposition from state regulators. At one point, in 2000, the Federal Energy Regulatory Commission proposed imposing a standard "market design" on electricity transmission operators across the US, but in the face of state protests had to withdraw the plan.

More US states might have deregulated of their own accord, had not the largest of them, California, made such an appalling mess of its own version of deregulation in 1999–2001. The state of California deregulated wholesale power prices – which rose by five hundred percent in a year – but failed to free retail prices as well or to let utility companies make any long-term forward hedges or insurance provision against future fuel-cost increases.

This pushed California's two largest utility companies into insolvency. Nor was the cause of deregulation helped by the Enron company's involvement in making money out of California's power shortages, although this had nothing to do with the scandal that eventually finished Enron off.

Margaret Thatcher's government brought a purer form of energy deregulation to the UK in the early 1980s. This was set out by her energy secretary, Nigel Lawson, in 1982 in classic *laissez-faire* terms. "I do not see the government's task as being to try to plan the future shape of energy production and consumption. It is not even primarily to try to balance UK demand and supply for energy. Our [government's] task is rather to set a framework which will ensure that the market operates in the energy sector with a minimum of distortion and energy is produced and consumed efficiently." The framework Lawson spoke of was vigorous encouragement of competition that led to privatization and dismemberment of the UK's gas and electricity monopolies.

At the EU level, the UK, Scandinavia and the Netherlands pushed for opening up the European energy market to cross-border competition. However, the European Commission has, over the past dozen years, carried out its own crusade to try to forge a more integrated and more competitive energy market. At the Commission's urging, EU law-makers have passed three successive pieces of legislation – in 1996–98, 2003 and 2009 – to give energy networks the necessary independence to act as common carriers of energy.

Regulation makes a rebound

The "free market" ethos behind these trends of energy liberalization and globalization has suddenly found itself under assault in the form of governmental calls for more regulation of oil-market speculation. This came after the extraordinary see-sawing of the oil price in 2008–09: up to $147 a barrel in July 2008, down to $30 by December 2008 and then back up to $70 by mid-summer 2009. Such volatility makes both financial and energy planning for governments very difficult. Despite the relative steadiness of the oil price since, by 2010 both the US and the European Union were looking at the possibility of reducing the volume and increasing the transparency of speculative trading in energy commodities.

The oil-price instability also prompted a flurry of government targets for increases in renewable energy, and for reductions in unnecessary energy-use and greenhouse-gas emissions. The outbreak of target-setting has been greatest in the UK, possibly as compensation for its own poor performance in these areas. Britain was already committed, through its European Union membership, to the EU-wide goals of a twenty percent emission cut and of a twenty percent average renewable share of total energy use by 2020. The UK role in achieving the EU average renewable target commits Britain to the biggest stretch of a target in renewable energy of any of the 27 EU states. Moreover, the UK has gone ahead and passed its 2008 Climate Change Act. This goes even further than the EU commitment on emissions. The Act imposes on future governments "legally binding" targets of a 26 percent emission reduction cut by 2020, and a series of legally binding five-year national carbon budgets for the country. (It is not entirely clear what kind of legal redress could be taken, or by whom, were the UK to fail to meet its targets.) The US has subjected itself to somewhat fewer targets than the UK or the EU, but is beginning to catch up with its federal mandate on biofuels and the minimum renewable energy standards already set by 29 individual US states.

The reason for all this is obvious. Markets are useful, perhaps even essential, when it comes to matching supply and demand: today's oil markets have prevented – at a price – any of the shortages of the 1970s era recurring. But markets cannot take proper account of carbon until it has a price that truly reflects carbon's social welfare cost (as economists would say), or its damage to the environment. Governmental regulation, subsidy and intervention are therefore needed to produce the low-carbon energy outcomes that markets have so far not delivered.

Liberalization is under scrutiny, due to pressing concerns about energy security – which increasingly acknowledges stability of price as well as quantity – and of climate change. The state is stepping back into energy to deal with the market problem of oil prices and the market failure of climate change.

Oil prices: chronic instability?

"The problem of oil is that there is always too much or too little", as Paul Frankel said in his *Essentials of Petroleum* in 1946. Not only is oil a key commodity, but its price is also chronically unstable. Some would claim that oil prices are *inherently* unstable, because specific features in both the supply and demand of oil mean that supply does not rapidly self-adjust to changes in demand, and vice versa.

Oil and gas projects tend to be slow to set up and get going. They tend to be projects with a long lead-time to get the necessary commercial permission and to make the necessary technical preparations – which is particularly problematic if they are located offshore. Once started, they can be slow to stop – in response, say, to a fall in demand – for geological and commercial reasons. Oil is almost *physically* prone to boom and bust.

New oil fields tend to flow rapidly and easily to begin with, whereas solid minerals like coal have to be hewed out of the ground at a constant rate. Oil executives are also keen to see a speedy return on high exploration and development costs. For instance, much of the colossal $136bn estimated total cost of developing the huge Kashagan oil field in the north Caspian sea will be spent in the fourteen years between its discovery (back in 2000) and the expected delivery of its first oil in 2014. Relative to this, the cost of actually exploiting fields and pumping oil (or gas) can be fairly insubstantial. So there is every incentive to keep pumping regardless of ups and down in demand.

Another reason for instability in the oil price is that consumption of oil is not very sensitive to its price. Or as the economists would say, the demand for oil is relatively price-inelastic. The main factor here is the lack of substitutes for oil, especially as transport fuel. (Power generation, by contrast, has seen oil largely replaced by coal or gas throughout the world.) Of course, if petrol prices plummet, people drive more, and less

A worker at the Bolashak processing plant near the Kashagan offshore oil field in the Caspian sea, western Kazakhstan, in August 2009. Kashagan is the world's biggest oil discovery since 1968.

if petrol prices soar. But lack of any easy substitute for petrol, diesel or jet kerosene means that the demand for these crude oil-based fuels alters relatively little if the price changes. This is why European governments can get away with putting such high taxes on petrol: our mobility is simply too important to us to be radically affected by oil's price. Imagine what would happen if governments were to add a comparable tax of up to eighty percent on the price of a foodstuff. People would rapidly change their diet and switch to other foods.

At all events, there has been a tradition, for almost the entire 150-year life history of the oil industry, of maintaining price by regulating production. Ironically, considering the fact that US governments have often led the complaints about oil cartels, this tradition of oil cartels started with the Texas Railroad Commission and its equivalent body in Oklahoma: they imposed production quotas on oil producers to prevent them undermining their own livelihoods by pumping too much oil and driving the price of their product to rock-bottom. Internationally, this market regulation role was taken over by the multinational oil companies, the famous Seven Sisters, some of whom survive in today's oil big boys – ExxonMobil, Shell, BP and Chevron.

Until the early 1970s these multinational companies dominated the oil market (see p.151). They agreed on prices known as posted prices, which were set to determine the income, or royalties, that the governments of oil-producing nations would receive. These posted prices did not respond to forces of supply and demand, because they did not have to. Oil trading, such as it was, was really a form of inter-company exchange between the Seven Sisters. Then, in the early 1970s, came the revolt of OPEC.

OPEC's rise

More than a decade after it was formed in 1960, the OPEC countries decided to take advantage of the fact that a tight oil market was giving them increased pricing power. The fourth Arab–Israeli war (also known as the Yom Kippur war) broke out in 1973, which exacerbated tensions between Middle East oil producers and the West in general – and the US in particular. OPEC's share of world oil production had risen to 52 percent.

Taking the initiative on posted prices for the first time, the six Gulf members of OPEC unilaterally announced successive rises in the posted price of Arabian Light crude oil from $3.65 to $11.65 per barrel during the course of that year.

At the same time, OPEC's Arab member states formed OAPEC (the Organization of Arab Petroleum Exporting Companies) and imposed a politically motivated oil embargo against the US and other countries supportive of Israel in the war. Needless to say, the resulting oil shortages provoked crises throughout industrialized nations that relied on OPEC for their crude.

The 1973 oil crisis hit the US hard, with severe shortages at gas stations and a rota system (based on the numbers on car numberplates) introduced.

OPEC and the oil market

For the next dozen years, the basis of the pricing system was a "marker" price for Arabian Light, effectively set by Saudi Arabia (being the dominant producer), as the benchmark for all other crude oil prices. But the system gradually broke down, largely because trading increased.

This was because many OPEC governments nationalized the upstream (exploration and production) reserves of international oil companies, which left companies increasingly short of enough crude for their downstream refineries. So the international companies had to look for supplies elsewhere. By the late 1970s and early 1980s they were beginning to find this new non-OPEC oil in areas such as the North Sea, the Soviet Union and Mexico, and from suppliers ready and able to undercut OPEC prices. As a result, OPEC's market share shrank from 52 percent in 1973 to less than 30 percent by 1985. Saudi Arabia's attempt in 1985–86 to recover market share only succeeded in slashing the oil price to less than $10 by 1986, causing a complete breakdown of the system of administered prices.

A system of market-related prices arose out of the final failure of attempts to fix oil prices. Three main price benchmarks are used, and the prices of other crudes are set in relation to them (usually much lower, because the benchmarks are chosen for being relatively light and low-sulphur). These benchmarks are: "West Texas Intermediate" (WTI) for sales to the US; "Brent", a North Sea blend, for sales in Europe and some other areas; and a crude called "Dubai" for most sales to Asia and the Middle East.

In theory, the benchmark price is the "spot price" of physical cargos traded on any one day. The spot price is the price quoted for immediate payment and delivery. But the volume of at least Brent and Dubai is shrinking, giving rise to fewer cargo shipments, fewer daily sales, and therefore prices that may be unreliable. So the effective benchmarks have become the prices derived from the far bigger futures markets – agreements between two parties to buy or sell an asset at a given point in the future – for these three crudes. Here liquidity, meaning the constant ability to buy, sell and trade, is guaranteed by speculators taking positions to make money, and a proliferation of producers, refiners and consumers hedging their bets on price. The volume of "financial" oil traded every day has now increased to many times that of the "physical" oil.

If OPEC didn't exist, would we have to invent something similar?

The short answer: yes, we probably would. The inherent instability of oil prices requires some effort to smooth out their peaks and troughs. OPEC does not try to fix the oil price directly (at least not since 1986), but instead tries to influence its level by varying production – a practice that had a precedent among producers when international commodity agreements used to exist governing sugar, tin, coffee and rubber production. Eleven of OPEC's oil-exporting member countries – all except Iraq, in fact, which does not currently participate in the production quota system – routinely cut their output to raise prices and increase their output to lower prices.

Particularly when criticized for high oil prices, OPEC often claims to be a price-taker rather than price-setter – in other words, it has to take as benchmarks the prices set by the future markets for WTI, Brent and Dubai (see previous piece). This will strike many as a joke. But it is true that the cartel only has indirect influence on price through the output

Saudi Arabia's oil minister Ahmed Zaki Yamani (centre), and members of his delegation, at the opening ceremonies of the 59th OPEC Conference at Kuta Beach, Indonesia in December 1980.

At this conference, oil ministers from the 13 members of OPEC agreed to permit crude oil prices to rise to a new high of $41 a barrel. (Yamani is now best known than the oft-quoted observation he made many years later – that "the stone age did not end for lack of stone".)

quotas it sets for its members. OPEC production changes can take up to three months to take actual effect, because of time-lags involved in ramping up production or shutting down wells and in shipping oil overseas. However, OPEC quota decisions signal to the market what OPEC's preferred range of prices is, though there is nothing precise about the outcome. The futures markets will then take a view as to the credibility (or lack of credibility) of OPEC announcements to cut or increase production and will therefore mark oil up or down.

The spare capacity weapon

OPEC's ability to influence the oil price depends essentially on its willingness to hold production capacity in reserve when prices and demand are weak, as well as to use it when prices and demand pick up. Keeping capacity idle is not a normal thing for anyone to do. Even when demand for oil slumps, non-OPEC oil producers do not reduce output. This makes it doubly hard for OPEC. When the cartel reduces output to maintain prices, it has to watch non-OPEC producers take its market share and, free-riding on the cartel's self-restraint, reap the benefit of higher prices.

This is why OPEC members so often cheat and pump more than their alloted quotas. But it is spare capacity, and the willingness to increase that spare capacity if necessary, that gives OPEC some influence over the market. In practice, this influence comes almost entirely from OPEC's dominant member, Saudi Arabia. By raising total capacity to 12.5m barrels per day (mb/d), Saudi Arabia's spare capacity exceeds the production of the next biggest producer, Iran (see table below).

OPEC country production, November 2009

OPEC member	Output (in millions of barrels per day)
Algeria	1.25
Angola	1.89
Ecuador	0.49
Iran	3.73
Kuwait	2.27
Libya	1.52
Nigeria	1.99
Qatar	0.8
Saudi Arabia	8.25
United Arab Emirates	2.10
Venezuela	2.2
Iraq	2.43

Source: Centre for Global Energy Studies

OPEC claims to want to keep oil prices just like Goldilocks' porridge – not too hot and not too cold. In practice, however, OPEC likes its oil price pretty high. The oil producers would not be human if they did not take a more relaxed view of rising oil prices than of falling prices. So for most of the 2004–08 rise in the oil price to the record of $147 in July 2008, OPEC stayed pretty passive. Indeed, its only major move during this period was to cut supply in early 2007, because it was worried about a rapid build-up of oil stocks.

In the latter stages of this climb in oil prices, some OPEC members began to worry about high prices leading to "demand destruction" for their product. But that was nothing like the collective panic that led OPEC in December 2008 – as the extent of the financial crisis and economic recession became clear – to announce a 4.2mb/d cut back from the collective 29mb/d level they had agreed in September 2008.

A reduction of this size taxes OPEC discipline sorely. This was evident in spring 2009 even from OPEC's own publications (which, bizarrely, use secondary sources on the production of OPEC countries that generally refuse to divulge their own figures on their own output). This shows Iran, Nigeria, Angola and Venezuela to be the biggest cheaters on their output quotas. This was no surprise. Iran, Nigeria, and Venezuela all have big populations and all traditionally find it harder to accept quota cuts than the less populous states of the Arabian peninsula. In addition to the internal OPEC tension between large and small population states, there is also disagreement inside the cartel over the size of quotas which do not reflect reserves.

Certainly OPEC could be made more efficient. Its quotas could be made more rational. And it would help the cartel if its members did not treat their oil-production figures as state secrets. But perhaps the world should be grateful that OPEC is not more disciplined in its quota decisions. Such decisions, instead of smoothing oil-price movements, can sometimes seriously aggravate them. The classic case was OPEC's 1997 decision to ignore the effect of the Asian crisis of that year and to increase oil production, in a way that triggered a collapse of the oil price in 1998. The decision to ignore the Asian downturn was all the stranger given the fact that OPEC took the decision in Jakarta, one of the very few times that OPEC has met in Asia.

Climate change

The greatest market failure of all

Climate change is the common thread running through this book. With every year that passes, there are fewer and fewer desirable changes to today's energy system that are not related to climate change in some way. You could point, for instance, to the need to provide the poor in developing countries with modern fossil fuels for cooking and heating, so that they don't have to scour their environment for firewood that kills them with smoke-related diseases.

But even here, the pressing humanitarian concern is ultimately not just one of public health: the bigger picture would take in damage currently done by deforestation, which is a major cause of climate change. Adopting a pragmatic attitude to the sheer scale of the climate-change problem requires many of our legitimate concerns about certain energy sources to be reassessed.

Nuclear power is the most glaringly obvious example: the difficulty of disposing of nuclear waste safely, securely and sustainably has made it the elephant in the global-energy room. For all its physical risks and economic drawbacks, nuclear power has the great offsetting advantage of being the one non-carbon energy source capable of large-scale energy generation (see pp.85–99). The bottom-line assessment about any energy source these days will always come down to the question of whether or not it contributes to the build-up of greenhouse gases.

Climate basics

This book will not enter into the science of climate change in great depth – there is a list of recommended books given in the Resources section, among them the *Rough Guide to Climate Change*. But in order to relate climate change to the energy crisis, a few very basic points need to be recapped.

The sun's rays pass through the Earth's atmosphere and warm the planet. These rays would then escape almost entirely back into outer space, leaving us freezing at night, or whenever we were out of direct sunshine, were it not for greenhouse gases (GHGs), which trap some of the infrared radiation in, acting as a kind of blanket. So the greenhouse effect is a natural phenomenon, which happens to maintain Earth's surface at around 30°C warmer than it would otherwise be. This gives Earth a Goldilocks temperature – not too hot and not too cold. Mars has no GHGs and is freezing; Venus is a seething mass of GHGs and is boiling.

Most greenhouse gases are products both of the natural ecosystem and of mankind (thereby giving rise to the ongoing argument over what the precise human contribution to global warming is). The main GHGs are water vapour, carbon dioxide, methane and nitrous oxide. Overall, these greenhouse gases make up a very small part of earth's atmosphere, which is mainly composed of nitrogen and oxygen, and which interposes no barrier to heat escaping the planet.

The biggest greenhouse gas by volume is carbon dioxide, and so the other gases tend to be measured for their greenhouse-warming effect in terms of carbon dioxide equivalent, or CO_2e. When they are measured as a ratio of the whole atmosphere, they come out small – today's level of CO_2e is around 435ppm (parts per million), or ppm, of the atmosphere. But with regard to climate change, small numbers make a big difference. And this number is increasing: it has already increased significantly during the period for which we have records. So the blanket is getting thicker, and those under it – the Earth and all that dwell on it – are getting hotter.

Carbon dioxide is by far the most voluminous of greenhouse gases, amounting to 75 percent of them, and the burning of hydrocarbons in energy production is by far the biggest contributor to it. According to the 2007 report of the Intergovernmental Panel on Climate Change (IPCC) energy production accounted for nearly two-thirds of global greenhouses gases in terms of CO_2e. This figure is a global average and it is important to note the considerable difference in the contributions to it from industrialized and agricultural economies. In the US, energy-related CO_2

emissions (from burning oil, natural gas and coal) represent just over eighty percent of the country's man-made greenhouse gas emissions. In New Zealand the energy sector only contributes about 43 percent of total emissions: there, agriculture accounts for 48 percent of the country's total. Indeed, the largest single source of New Zealand's emissions takes the form of methane from the stomach belches of the country's large population of ruminant animals, predominantly sheep. Since these animals have been bred by New Zealand's livestock industry, these emissions count as man-made or, as the Greek-origin word has it, anthropogenic.

The IPCC and the Fourth Assessment Report

The Intergovernmental Panel on Climate Change is the scientific body, created under the United Nations Framework Convention on Climate Change, that monitors global warming. It collates and assesses information from a global network of experts, and is the most important information source in determining global policy on climate change. The IPCC published its Fourth Assessment Report in 2007 and its findings were alarming, to say the least. "The global atmospheric concentration of carbon dioxide has increased from a pre-industrial value of about 280ppm to 379ppm in 2005. The atmospheric concentration of carbon dioxide in 2005 exceeds by far the natural range over the last 650,000 years (180 to 300ppm) as determined from ice cores. The annual carbon dioxide concentration growth rate was larger during the last 10 years (1995–2005 average: 1.9ppm per year) than it has been since the beginning of continuous direct atmospheric measurements (1960–2005 average: 1.4ppm

The Kaya Identity

This is not the title of a thriller, but an equation written by a Japanese economist, Yiochi Kaya. It neatly summarizes the factors driving the world's man-made carbon dioxide emissions. Here it is:

$$CO_2 = Output\ (population \times GDP) \times Energy\ Intensity\ (energy\ use/GDP) \times Carbon\ Intensity\ (CO_2/energy\ use)$$

What it highlights is the difficulty in slowing emissions. They will go on rising because of population growth and income growth – two things no politician would dare to campaign against – unless, that is, there is an offsetting reduction in energy intensity and/or in carbon intensity. Our best hope for reducing energy intensity lies in radically increasing our energy efficiency, while the best chance of reducing carbon intensity lies in new technology. Both of these aims require a subtantial amount of political will and international cooperation.

IPCC Critics and climate sceptics

The IPCC has its critics. Some accuse it, out of hand, of exaggerating global warming. Others accuse the IPCC of producing reports that are too cautious and too infrequent to keep up with the pace of climate change. The process by which it produces its formal assessment reports on the state of climate change – of which there have been four (1990, 1995, 2001 and 2007) – is curious. Its working groups are a mix of scientists and government officials. The IPCC claims that "because of its scientific and intergovernmental nature, the IPCC embodies a unique opportunity to provide rigorous and balanced scientific information to decision makers," going on to confidently assert that "by endorsing the IPCC reports, governments acknowledge the authority of their scientific content" and that "the work of the organization is therefore policy-relevant and yet policy-neutral, never policy-prescriptive". Nonetheless, it is a strange, hybrid process in which the evidence comes from the scientists, but the conclusions to be drawn from it, contained in "the summaries for policy makers", are agreed between the scientists and government officials.

The Fourth Assessment Report (AR4) of 2007 has also been criticized by among others Joseph Romm, a former senior US Department of Energy official in the Clinton Administration, of underestimating the potential positive-feedback effects of climate change. There is a possibility that the Earth's rising temperature could release huge amounts of methane through the thawing of the Arctic tundra which currently keeps it buried: it is but one of many vicious-circle scenarios in which a rising temperature precipitates events that, in turn, increase the world's temperature even further.

James Hansen, head of the Goddard Institute for Space Studies in New York, has suggested that the IPCC has underestimated the degree to which the world's ice-caps are liable to melt – and, in turn, that they have underestimated sea-level rise. Hansen criticized the IPCC's 2007 report for shrinking back, on this occasion, from trying to evaluate possible dynamic responses of ice-sheets to global warming, while at the same time issuing some very precise numbers for estimated sea-level rises in the 21st century. Its full-range estimate is a rise of between 18 and 59 cm and its mid-range estimate is between 20 and 43. Hansen has complained that "the provision of such specific numbers for sea level rise encourages a predictable public response that the projected sea level change is moderate, and smaller than in the IPCC 2001 (AR3) report". He pointed out that it has led to "numerous media reports of 'reduced' sea-level rise predictions".

Some scientists who have contributed evidence to the IPCC have expressed reservations about its emphases. Dr Chris Landsea, a US hurricane meteorologist who had been involved in the IPCC's Third Assessment Report (2001), pulled out of the Fourth Assessment in 2005, complaining that his peers were trying to make too close a link between the frequency of hurricanes and climate change. (Significantly, however, he had no qualms in ascribing oceanic and atmospheric warming over the last decades to the increases in GHG.)

per year) although there is year-to-year variability in growth rates." The IPCC forecasted that, on present trends, this century would see a rise in global average temperature of between 1.8°C and 4°C by 2100.

The stabilization challenge

The global average temperature rise has already risen by nearly one degree centigrade above the preindustrial level. And, even if from today onwards no further GHGs were emitted, the momentum of emission increases in the recent past is

Professor James Hansen: one of several scientists who have criticized the IPCC's reports for not taking sufficient account of the extent to which climate change precipitates further climate change.

bound to carry us on to a two-degree increase. The aim, however, is to stop the warming there – to keep the temperature increase to no more than two degrees. Recognition of the need to confine the temperature rise to this point was almost the only element of agreement at the otherwise dismal United Nations climate-change summit in Copenhagen in December 2009. The pressing issue of how to achieve this, however, was not something that the 190 countries were able to agree there.

To keep within range of the two-degree limitation target, the general scientific consensus is that GHG levels ought to be stabilized at 450–550ppm of CO_2e. The major review of the economics of climate change, led by UK economist Sir Nicholas Stern (see p.41), took as its starting point a 450–550ppm target. Stern warned that "it would already be very difficult and costly to aim to stabilise at 450ppm CO_2e. If we delay, the opportunity to stabilize at 500–550ppm CO_2e may slip away." Stern was writing in 2006.

Current international climate-change negotiations are very much focused on reductions in the annual flow of new emissions. With the advent of the Obama administration, the US has now joined other developed countries in agreeing to aim for an eighty percent cut in annual emissions by 2050, as was agreed at the July 2009 Group of Eight summit. But it is worth remembering that it is only these G8 countries of

Opinion differs as to who is more to blame for the disappointing inconclusiveness of the Copenhagen conference in 2009. It was China's delegate who insisted that a target of an eighty percent cut in carbon emissions by 2050 for the rich, industrialized nations be dropped from the deal. (China would not have even been included in this target.) China also objected to the target of 2020 as the year in which global emissions peak.

On the other hand, Barack Obama faced criticism for brokering the eventual non-binding deal amongst a coalition that did not include, or negotiate with, poorer nations. Obama was also criticized for pushing for concessions from China while offering little, if anything, from the US.

North America, Europe and Japan that are even going to pay lip service to going as far as an eighty percent emissions reduction. These nations are the principal economic beneficiaries of the industrial revolution that caused the GHG build-up. Developing countries will need some highly convincing incentives to do likewise. China has recently overtaken the US as the world's biggest GHG emitter: it is currently the biggest carbon culprit, but as regards the current stock of GHGs that has built up in the atmosphere over the period 1850–2000, China is responsible for only six percent of emissions, compared to thirty percent for Europe.

An unflinching, rigorous analysis of the science would suggest that the entire world's annual amount of new emissions is going to have to be cut by *more than eighty percent* below today's level for emission levels, and therefore the climate, to stabilize. Only this would bring emissions down to less than five gigatonnes (Gt) of CO_2e a year – the level that the planet's soils, vegetation and oceans could naturally absorb. Earth's absorptive capacity is not inconsiderable. The Stern report suggested that of the 2000Gt of CO_2 released into the atmosphere over the past two centuries through human activities – chiefly the burning of fossil fuels, ploughing land and cutting trees – sixty percent had been absorbed in the land and sea, leaving 800Gts of CO_2 to accumulate in the atmosphere.

Getting a move on

There is a serious time-pressure for action. The longer the delay in cutting the flow of new emissions, the higher the stock of emissions will rise, and with it the temperature. "Every ten-year delay in achieving peak emissions adds another 0.5°C to the most likely temperature rise", says Vicky Pope of the UK Met Office. And, furthermore, it's important to remember that peak emissions refers only to the point at which a decline in emissions begins, not the point at which an eighty percent cut is achieved.

Unfortunately, there is evidence that rising temperatures weaken the absorptive capacity of carbon sinks – the aforementioned soils, vegetation, trees and oceans – mainly due to the carbon-absorbing vegetation dying off during droughts. For instance, the very hot summer of 2003 in Europe is believed to have turned the continent's terrestrial ecosystem from a carbon sink into a carbon source. The CarboEurope research project, funded by the European Commission, estimated that during July and August 2003, 500m tonnes of carbon were released into the atmosphere by Europe's fields and forests, around twice as much as the emissions from fossil-fuel burning in Europe during the same period.

For the energy sector, the cost of climate action delay could be huge. In 2009 the International Energy Agency estimated that every year the world delays shifting to policies that would keep emissions to 450ppm of CO_2e – which is now a highly ambitious target – adds \$500bn to the global estimated cost of \$10.5 trillion for mitigating climate change. This is because of the inertia in the world's energy-supply system. For instance, each year of climate-change inaction is another year for new long-lived, coal-fired, power plants to open around the world.

The Stern Review

The achievement of Nicholas Stern's influential report, "The Economics of Climate Change", was to show that we, as a world, need not bankrupt ourselves in fighting climate change provided we take early action, but that delay could be disastrous and costly. Specifically, Stern said that if early action were taken "to avoid the worst impacts of climate change", the costs "can be limited to around one percent of global GDP each year". Stern later doubled his estimate of mitigation of climate change to two percent of GDP, on the ground that climate change was accelerating and that it was therefore necessary to hit the harder target of keeping CO_2e concentrations below 500ppm, not just within the 500–550ppm range.

If the world carried on its business as usual, however, "the overall costs and risks of climate change will be the equivalent of losing at least five per-cent of global GDP each year, now and forever". Futhermore, "if a wider range of risks and impacts is taken into account, the estimates of damage could rise to twenty percent of GDP or more".

This could be "on a scale similar to those [disruptions] associated with the great wars and the economic depression of the first half of the twen-tieth century". Stern made headlines with his cost/benefit calculation that spending 1–2 percent of world GDP in early climate action could head off eventual climate mitigation and adaptation costs amounting to 5–20 percent of world GDP.

He was, however, widely criticized for using language and methodol-ogy that made the economic equation more lop-sided in favour of early climate action than seemed reasonable. He eschewed the traditional convention in welfare economics, which values a future benefit less than a present benefit, or which worries about a future cost less than a present cost. Rather, Stern said that assessing impacts over a distant future neces-sitates taking future generations more into account than ever before. "If a future generation will be present, we suppose it has the same claim on our ethical attention as the current one. Thus, while we do allow, for example, for the possibility that, say, a meteorite might obliterate the world, and for the possibility that future generations might be richer (or poorer), we treat the welfare of future generations on a par with our own."

In general terms, treating present and future generations "on a par" is fine. It squares with the ethical basis of sustainable development that most people today would accept. Namely, that future generations have a right to a standard of living no worse than the present generation's, and that the present generation, in fulfilling its own needs, has a duty not to compromise the standard of living of future generations.

But the Stern Review goes further, when it turns its generational parity approach into numbers. By putting present and future on an equal footing, a future eventual cost of unbridled climate change of "at least five percent of global GDP" somehow also becomes "the equivalent of at least 5 percent of global GDP each year, *now* and forever". This implication – of possible large and immediate climate costs – is not borne out elsewhere in the Stern Review. Even its high range of damage estimates shows costs only building up gradually, and still to under three percent of world GDP by 2100. This doesn't sound quite like a repeat of great wars or depressions – at least not in this century.

It is important to remember that Stern's report dates from 2006 – with each year that passes, it becomes harder and harder for the words "early action" to mean anything. Furthermore, climate-change modelling is not an exact science: some of the positive-feedback scenarios posited by climatologists could lead to exponential, runaway climate change that would radically alter Stern's climate-change accountancy.

But even at the time, no one queried Stern's stress on the energy sector as the main focus for climate action. The report called for the power sector around the world to be "de-carbonized" by at least sixty percent by 2050, and also urged deep cuts in transport emissions. Stern called climate change "the greatest market failure the world has ever seen" – particularly because the considerable damage of carbon to society and to the environment in general was not, and still isn't, reflected in the pricing of the energy producing the carbon. So, in addition to prescriptions for energy innovation and efficiency, the Stern Review lent its weight to a proper pricing of carbon, by whatever means: tax, trading or regulation.

Energy and climate change; climate change and energy

This book focuses on energy's impact on climate change and vice versa. Energy production is climate change's main cause, and the transition to a low-carbon energy economy is a necessary one. But the changing climate will have other impacts on energy – less remarkable perhaps, but also less remarked upon.

For climate change threatens existing methods of energy production in very real, physical ways. Even if we were short-sighted enough to determine to keep our current energy-production methods just the way they are, we would still have to acknowledge how vulnerable they are to the extreme weather that climate change is likely to send our way.

Across all Arctic zones and the tops of mountain ranges, the melting permafrost will cease to be a reliable foundation for energy infrastructure. In other words, the existing oil and gas pipelines are vulnerable, and it will become necessary to rebuild them entirely or in part; it will certainly become harder to move heavy equipment across any Arctic tundra that is thawing. This is obviously a problem for the main Arctic coastal states – the US, Canada and Russia.

Climate change will also alter wind resources, but it is impossible to say how, because wind depends on the differentials in temperature and solar radiation – not on any *precise* level of them. It is also likely to see an increase in extreme weather and storms, which will have immediate effects on offshore energy facilities in places such as the Gulf of Mexico. The impact of climate change on water supplies will have an immediate effect on hydroelectric plants especially those mainly dependent on glacial melt as in the Himalayas, Andes and Alps or on seasonal rainfall (such as the Indian monsoon) which may become less predictable.

Water shortages will also affect the cooling of conventional thermal and nuclear power stations, which are usually placed on coasts or rivers to draw water for cooling. Spain's prime minister has given shortages of water for cooling reactors as one of his reasons for not reversing his country's phasing-out of nuclear power.

Countries such as the UK which have placed virtually all their nuclear reactors on the coast face an opposite threat – too much water from rising sea levels. The UK Met Office has done some scenario planning which shows that by 2100 there could be increases in sea-surge heights ranging from up to 1.7 metres at Sizewell on the Suffolk coast, the most affected site, to 0.9 metres at Hinkley Point in Somerset, the least affected site.

It should be clear, then, that in terms of energy-generation – let alone on humanitarian grounds – business as usual is not an option.

PART 2

THE ENERGY GAME BOARD

Different fuels and the
parts they play

Oil

The black stuff that keeps the world moving

Oil is the mainstay of our energy system. But it is so much more because it is also, like natural gas, the raw material for making the many plastics and synthetic materials that we use at home, in offices and in factories – and the things we buy or wear every day.

The basics

In common with natural gas, which is just a lighter form of petroleum product, oil is the result of zooplankton, algae and bacteria being slowly cooked for centuries, without oxygen, under layers of mud and rocks at high temperature and pressure. Some oil and gas is probably still being formed today, albeit at several kilometres below the earth's crust where the temperature is hotter.

But most available oil and gas dates from long ago. For instance, North Sea oil is mainly found in rocks from the Jurassic period of 150 million years ago. Seas and swamps were then rich in microscopic plants and animals. When these died, they sank to the bottom, forming thick layers of organic material that became trapped in layers of mud. As the mud became thicker and hardened into rock, it began to act as a pressure cooker on this organic material.

Oil comes in many forms. They range from the lighter grades, which are relatively easy to refine, the heavy oil plentiful in Venezuela through to the sticky bitumen-like tar mixed in with the sands of western Canada. The Canadian and Venezuelan deposits are enormous, but are called unconventional oil because they need special heating and treating in order to get them out of the ground and flowing along pipes. Conventional oil is more fluid. At the start of an oilfield's life, it is often sufficient just to drill into the oil field for the oil to flow out under its own pressure. Later on it is usually necessary to inject, or re-inject, gas or water into the field to keep the pressure up and, therefore, the oil flowing. Traditionally, oil wells have been drilled straight down.

But increasingly drillers branch out horizontally or sideways out of the main oil wells, in order to reach more pockets of oil without littering the land or seascape with drilling rigs. This form of drilling, which is also known as directional drilling, has been made possible by the development of steerable rotary drills. It can be done for as far as ten kilometres, as BP has shown by drilling under Poole harbour from its Wytch Farm oil well on the English south coast. In 1990 Iraq claimed that neighbouring Kuwait was using slant drilling to steal Iraqi oil, a claim it invoked to justify its invasion of Kuwait that year.

The top six oil producers in 2008 (millions of barrels a day)

Saudi Arabia	10.8
Russia	9.88
United States	6.73
Iran	4.32
China	3.79
Canada	3.23
World production*	81.8

Source: *BP Statistical Review*

The top six oil consumers in 2008 (millions of barrels a day)

United States	19.4
China	7.99
Japan	4.84
India	2.88
Russia	2.79
Germany	2.5
World consumption*	84.4

Source: *BP Statistical Review*

*Consumption was slightly higher than production because of stock changes, non-petroleum additives and substitute fuels.

The future of oil
Why we should welcome a peak in oil demand, but fear a peak in oil supply

Oil remains the biggest primary energy source. (Electricity has to be derived from something else and so is a secondary energy source.) Of global primary energy demand in 2006, oil accounted for 34 percent, coal 26 percent and gas 21 percent. More than half (52 percent) of world oil now goes into transport: oil fuels more than 90 percent of the world's transport needs. While transport's demand for oil keeps rising, especially in developing countries, the share of world electricity generated with oil has halved in the last thirty years to just six percent in 2006, again mainly in developing countries. So the broad picture is that oil demand would seem to have peaked in richer, industrialized countries, which have had some success in weaning their power sectors off oil reliance, and are struggling to do the same with their transport sectors.

But it is too early to speak of any peak in *overall* oil demand. Demand is still rising among some major developing countries that are populous (India), rich (Saudi Arabia) or both (China). The International Energy Agency forecasts that over eighty percent of the increase in oil demand to 2030 will come from China, India and the Middle East. It is a pity that a peak in oil demand still looks some years off: a peak in oil demand would match the peak we are seeing in the supply of easy-to-find and easy-to-extract oil (outside of the huge reserves controlled by OPEC producers). Less demand for oil would keep the price from soaring as supply becomes scarcer. Ever-increasing world oil demand will have just the opposite effect, and could one day make the mid-2008 price rise to $147 a barrel look cheap.

Peaks of supply and demand

Oil supply will probably peak before oil demand. In many parts of the world it is happening already, as shown by the chart overleaf. Why does the energy world spend so much time discussing "peak oil", when "peak gas" or "peak coal" go undebated? Well, there are three good reasons why we would definitely not want a runaway rise in the oil price – which would be triggered if world oil supply goes into a clear decline.

The first is that everyone is affected by the oil price, because oil is still the world's main primary energy source. There are few substitutes yet

Non-OPEC countries in decline or plateau

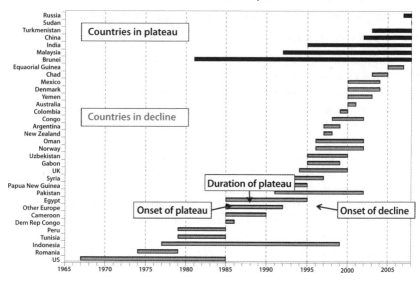

Source: PFC Energy, *Global Liquids Supply Forecast*

The bars show the onset and duration of documented production peaks or plateaus – tracking country life cycle shows an acceleration of the number of countries passing from peak to decline.

for oil in transport, which is over ninety percent oil-derived. Oil is the only fuel to so dominate any sector: elsewhere energy sources or fuels are largely interchangeable via the medium of electricity. But the only major, large-scale transport application of electricity so far has been to trains.

Just as everyone is affected by the oil price, the oil price can be affected by oil-supply (and demand) changes anywhere on the globe. Oil is the only unified global energy market. (In the case of energy sources, such as gas, there could be a supply shortage or excess in one region with hardly any impact on other regions.)

Finally, not only does the oil price *absorb* oil supply/demand shocks around the world – it also *sends out* shocks of its own. For instance, continental Europeans link their gas prices to the oil price: oil is thus the prime mover for European energy prices. The level of the oil price is a powerful determinant of world economic activity: the state of the global economy and the price of oil tend to move up and down together.

Peak oil supply: a matter of when, not if

For all these reasons, it is important for all who use or trade oil to know or try to guess how much of the black stuff is left. Obviously, oil will decline and run out one day: just like gas and coal, it was formed in the distant geological past and its supply is finite. The issue is when. (It is worth noting here that oil companies often take issue with statements about oil running out: they prefer to frame the issue in terms of the point at which it becomes "uneconomic" to access "difficult oil".) The level of oil output could be maintained, perhaps even for a long time, by turning to unconventional oil – the particularly sticky or viscous oil found in the oil sands of Canada, shale rock in the US Rockies and in the Orinoco belt of Venezuela. The cost of extracting this unconventional oil is high – and it's not just a financial one. The difficulty of access means more despoliation of the landscape, and the carbon footprint of the extra processing required is a significant environmental minus, to say the least. And the day the world becomes dependent on unconventional oil to meet any increase in demand will be the start of a sustained price rise.

The International Energy Agency tells us that, in the 150 years since the modern oil industry began in 1859 in Pennsylvania, the world has probably used around 1.1 trillion (thousand billion) barrels of conventional oil. This would tell us how much is left, if we knew what the world's total endowment of oil was in the first place. But of course we don't. And it is important to realize that the volume of remaining, reasonably recoverable reserves of oil may be essentially unknowable. It will depend on levels of investment and prices, technological developments and, as ever, climate change. Increases in the frequency and ferocity of hurricanes, for instance, would make it harder to drill in the Gulf of Mexico.

On the other hand, global warming will open the Arctic ocean to more exploration for oil and gas. That the response to an unprecedented climatic change wrought by man-made global warming is to exploit it – in a way that will precipitate further climatic change wrought by man-made global warming – is not exactly a cause for celebration among environmentalists. There is already some onshore production along the coastal rim of the Arctic ocean in Alaska and Siberia, and we are beginning to witness offshore exploration off Greenland: it has put eight areas along its west coast out to tender, which were won by major international companies alongside Greenland's own national oil company.

The black art of estimating reserves

The business of estimating reserves begins with science. Oil explorers take seismic surveys, or acoustic reflections, of cross-sections of the earth's crust, both onshore and offshore. But while these may provide a promising picture of oil and gas deposits, they are only clues – not proof – of oil and gas deposits. Only drilling can provide hard evidence of the presence of oil or gas. This is why the US Securities and Exchange Commission insists on seeing drilling evidence to back up companies' reserve claims. As financial regulator of the world's first international oil companies, the SEC has traditionally led the way in setting standards and definitions of reserves for publicly-listed oil companies.

But just because an oil deposit exists doesn't mean it is extractable by a reasonable means or at a reasonable price. This is where objective science begins to give way to subjective judgement. On a series of assumptions which take into account marketability and future prices as well as cost, geology and technology, oil and gas reserves are divided into:

▶ **Proven reserves:** a ninety percent probability of being extracted profitably

▶ **Probable reserves:** a fifty percent likelihood of profitable extraction

▶ **Possible reserves:** a ten percent chance of profitable extraction

This requirement of a degree of commerciality for reserves is essential for investors to be able to gauge the practical value of the reserves held by publicly listed oil companies. Commerciality is subjective and can sometimes appear to exist only in the eye of the reserve-holder. In 2004, for instance, Royal Dutch Shell was fined by the US and UK regulatory authorities for deliberately overstating its reserves. To some extent, this was an attempt to make up on paper for reserves that the company had failed to find underground with the drillbit. But there were also issues of commerciality and marketability. The company was forced to take off its balance sheet some twenty percent of its reserves, including gas reserves from its Gorgon gasfield in Australia for which there was no firm buyer, and some oil in Nigeria whose export was constrained by that country's OPEC membership.

Calculations of the reserves of national oil companies, in particular those owned by OPEC governments, are even more arbitrary. Because they are not listed on stock exchanges, most of them do not require any outside audit of their reserve claims. For the same reason, the reserve claims of

M. King Hubbert and the Peak Oil controversy

Getting an oil supply forecast right for one country with excellent data is difficult enought. But it's far easier than attempting to do it for the entire world oil sector, in which much information is hidden, inadequate or non-existent. This is what the disciples of M. King Hubbert have found.

Hubbert was an American oil geologist who successfully predicted the peak in US oil production. In 1956 he examined oil-production statistics from the lower 48 US states, and plotted them into a bell-shaped curve. The chart projected that their production would rise for at least a dozen more years, but then reach a crest in the early 1970s. He was proven right: the peak arrived in 1970. Hubbert has been somewhat of an icon to conservationists ever since. But applying his predictive methods to world production has been difficult.

Oil fields typically share this bell-shaped production profile that Hubbert plotted. In contrast to the steady state of, say, coal production, output initially rises very sharply, as natural pressure pushes the oil out. But then it peaks and tails off as the volume of remaining oil – and therefore the pressure of it – diminishes, despite any enhanced recovery techniques.

If we could predict the peak, then we would have a better idea of remaining reserves – and vice versa. Knowing the timing of the peak in world oil production and/or the level of remaining reserves would be very useful. But apart from the general opacity of world oil data, there is the distorting factor of OPEC countries in the Middle East. Their production does not necessarily mirror the reserve profile of their oil fields: it is often deliberately reined in to sustain price.

In 2001, Kenneth Deffeyes, a Shell colleague of Hubbert's and later a Princeton University professor, used Hubbert's analytic methods to predict that world production would peak sometime between 2004 and 2008. It didn't. And Deffeyes and other "peakists" have been ridiculed by many in the oil industry.

A cheery-looking M. King Hubbert, the patron saint of peak oil.

The ridiculers argue that oil output can be almost indefinitely extended by exploiting unconventional oil deposits (disregarding the environment impact), by assuming that new technology will be developed (as it always has done in the past), and to some extent by profiting from global warming to open up the Arctic to exploration. In short, they accuse the peak-oil believers of crying wolf. Perhaps they should at least bear in mind that, in the story, the wolf eventually shows up.

national oil companies do not have to be backed up with drilling evidence (though they frequently are). Moreover, in the case of OPEC member countries, there has sometimes been a strong motive for governments to exaggerate their reserves.

For the size of OPEC members' production quotas within the cartel is supposed to be related to the size of their reserves. The biggest revision of reserves in history came in 1986–87 from OPEC countries, mainly in the Middle East. This revision was largely due to internecine OPEC politics – the re-negotiation of member countries' production quotas – and had little to do with discovery of new reserves or new appraisal work on discovered fields. The arbitrary nature of OPEC's reserve upgrades was highlighted by the fact that in the 1981–2003 period, stated reserves in OPEC fields increased by eighty percent overall, compared to a thirty percent reserve rise in non-OPEC fields.

So what's left?

The firmest estimate is, not surprisingly, the smallest. According to the International Energy Agency, at the end of 2007 another 1–1.1 trillion barrels of *proven* reserves of oil remained to be extracted. In other words, we are halfway through the oil – which sounds very much as though the peak oil moment is upon us.

But remember that "proven" is the category with a ninety percent chance of being extracted profitably. If you add in probable reserves (with fifty percent likelihood of profitable extraction), and allowance for normal (not OPEC-style) reserve growth due to reappraisal of existing fields, plus an estimate for recoverable oil from fields yet to be discovered, the IEA believes the world might have as much as 3.5 trillion barrels of "ultimately recoverable conventional resources".

This already huge number can be doubled to 6.5 trillion barrels, if you add in unconventional, heavy oil from the swamps of Venezuela, from the oil sands of Alberta and from the oil shale of the Rockies. This is what the IEA terms "the total, long-term potentially recoverable oil resource base". Nothing if not thorough, the agency says that the addition of coal-to-liquids and gas-to-liquids techniques could raise potential to nine trillion barrels. (These two techniques involve a German process – used during World War II – to produce synthetic liquid fuel out of coal or gas).

But all estimates of unconventional oil production are enormously dependent on financial and environmentat costs – which are already irrevocably intertwined and which will grow increasingly so as the twenty-first century progresses. Burning up the maximum estimated amount of conventional oil – 3.5 trillion barrels – might well make the planet uninhabitable for humans.

Plateau oil

Attempting precise predictions about future oil production is a mug's game (see box on M. King Hubbert and Peak Oil). So most future-watchers these days hedge their bets by forecasting that the long rise in oil output will start to level out in a few years' time, or resolve in a couple of decades into a plateau – a sort of multi-year horizon of peaks – before beginning its decline.

How high will the plateau be? That depends on how supply and demand factors play out. The increase in oil demand these days comes from the transport sector worldwide and from developing countries. If richer countries – in the name of sustainability – can start running their cars on biofuel or electricity, and if poorer countries can cut oil out of their power generation sectors altogether, then global oil demand would slacken. But the effect and timing of demand-reduction policies is hard to gauge. On the supply-side argument, one leading oil industry figure, Christophe de Margerie, the head of French oil major Total, has stuck his neck out to forecast that world oil production, currently around 85m b/d, will never go above 100m b/d. His main reason for saying this is that too much of the world's oil is closed off to international oil companies like his, and instead is in the hands of national oil companies with no commercial or political cause to maximize production. He has also been forthright in stating simply that the oil industry has been over-optimistic about the geology, "not in terms of reserves, but in terms of how

Christophe de Margerie, CEO of Total Oil, forecasts that, due to the declining quality of oilfields currently being exploited and the big question mark hanging over oil's "known unknowns" – Iraq, Venezuela and Nigeria – oil production will not increase to anything like the level that the IEA predicts.

Oil exploration technology

There have been many ingenious advances in oil exploration and production over the years:

▶ **Seismic technology** A process similar to the use of sonar waves and echolocation to detect objects underwater or in the air, seismic technology involves beaming energy waves at the earth's surface or seabed and constructing a picture of the subsurface from how these energy waves are reflected by the earth's crust. It is essential technology for oil geologists. They have successively developed two-dimensional (2d) seismic to create a picture of a narrow cross-section of the earth's crust, then 3d seismic to widen this cross-section, and even, in recent years, 4d seismic to add the dimension of time, indicating any likely movement or flow of liquids.

▶ **Horizontal drilling** This reaches a wide range of oil and gas pockets from one drilling, rig. It does not always refer to literally 90° drilling but simply to any wells that are more or less horizontal. It has a distinct advantage in that it can access oil resources impeded by surface obstructions, such as bodies of water or buildings. The US Department of Energy has indicated that horizontal drilling can lead to an increase in reserves of 2% of the original oil in place. The production ratio for horizontal wells versus vertical wells is 3.2 to 1, while the cost ratio of horizontal to vertical wells is only 2 to 1.

▶ **Enhanced oil recovery** The injection of water, natural gas, steam or liquid carbon dioxide into wells facilitates the extraction of oil from them by increasing the pressure within them. The natural pressure or gravity on a deposit will normally only push between five and fifteen percent of the oil out of the well. But if water or gas is pumped in, up to another thirty percent can be extracted. Reducing the viscosity or glueyness of the oil by pumping in steam can extract still more.

▶ **Deepwater drilling** Much of the growth in non-OPEC oil output, especially in the "golden triangle" of the Gulf of Mexico, off Brazil and off West Africa, has been achieved by drilling in deep water (between 400 and 1500 metres) and in ultra deepwater (deeper than 1500 metres). The advent of deepwater drilling has turned conventional wisdom about offshore oil exploitation on its head.

to develop those reserves, how much time it takes, how much realistically you need." Meanwhile, Sadad al-Huseini, the former head of exploration and production for Saudi Aramco, has provided an analysis showing that global oil production has already hit a plateau and is unlikely to rise above current levels.

In contrast, the IEA projects – on the basis of policies in force before the United Nations climate summit of December 2009 in Copenhagen – that global oil production will break through de Margerie's 100m b/d barrier. In its 2009 annual outlook, the IEA estimates that, on current trends, world oil output will rise from 84.6m b/d in 2008 to 88.4m b/d in 2015, to reach 105.2m b/d by 2030.

Running to stay still

New recovery methods have helped, but the gain in recovery rates barely compensates for an increasing rate of decline in existing fields. (Indeed enhanced recovery techniques may only accelerate the decline rate, enabling operators to suck out the oil quicker than before.) The reason for decline rates outpacing recovery gains is simple. The volume of oil in new discoveries fell behind the volume of production in the 1980s, and stayed there ever since. The new fields that are discovered tend to be smaller; earlier discoveries tended to be bigger fields because by definition they were easier to spot.

The IEA has found that smaller fields deplete quicker. In a massive survey of 580 of the world's largest fields that are past their peak, the agency found an average 5.1 percent annual rate of decline, ranging from 3.4 percent for "super-giant" fields to 10.4 percent for merely "large" fields. The cause of the differing decline rates may have more to do with policy than geology or technology. For decline rates are lowest in the Middle East, where national oil companies tend to want to husband national resources, and highest in North America where developers are often in a commercial hurry to recover their costs with rapid production.

An enormous amount of investment would be needed to offset these decline rates in oil supply – far more than would be needed to meet any conceivable increase in oil demand. To offset the decline rates, it is estimated that the world would have to bring on stream an extra 64m b/d of capacity between now and 2030 – roughly six times that of Saudi Arabia today – just to keep output steady.

Going offshore

Britain only has one sizeable onshore oil operation Wytch Farm and even that consists mainly of drilling horizontally to reach oil underneath Poole harbour in Dorest. So it had little choice but to go offshore in its search for oil. Over the past 41 years the UK has extracted nearly 39bn barrels of oil or oil equivalent (including gas), mainly from the North Sea, with its gas fields in the southern part and oil fields in the northern part. Oil and

Saudi Arabia's Ghawar oilfield – King of Kings

Ghawar is by far the world's biggest oil field, both in size 280km long and 30 km wide and in production. It has produced some 55 billion barrels of oil to date, and in 2007 was still producing 5.1m b/d – or 7 percent of the world's total conventional oil output. It also contains enough gas, alongside the oil, to be counted in the top ten gasfields of the world. Ghawar is probably the one field in the world where peaking and decline of production could, by itself, have a measurable effect on the world market and oil price. Saudi Aramco claims that peaking is some way off, because Ghawar still has "more than 70bn barrels of remaining reserves".

However, in his 2005 book *Twilight in the Desert*, US oil expert Matt Simmons cast doubt in general about the sustainability of Saudi production, and in particular about Ghawar's future output. Simmons suggested that, as they moved intensive exploitation of the field progressively down the field from north to south, Saudi Aramco engineers seemed to assume they would find a continuation of the same extraordinary productivity they found at the start. The US expert questioned the wisdom of that assumption.

The International Energy Agency's *World Energy Outlook* of 2007 did not include Ghawar among the "post-plateau" fields, because although Ghawar's production in 2007 was below its historical peak of 5.6 million barrels per day reached in 1980, it was less than 15 per cent below it. The observed post-peak decline rate is thus a mere 0.3 percent per year.

Gas UK, the industry association, reckons there are still 25bn barrels of oil equivalent potentially out there.

The UK has been criticized for getting through its oil and gas reserves faster than other North Sea coastal states, a point underscored by the present decline in production. Britain is no longer self-sufficient in either oil or gas. In 2008 it produced 1.54m barrels a day of the 1.7m b/d that it consumed, and 69.3 billion cubic metres of gas out of the 93.9bn cm that it consumed. But it can also be argued that the UK was bound to deplete its oil and gas quicker than Norway and the Netherlands, simply by virtue of having a larger population than those countries. At all events the North Sea has been an undeniable, if temporary, godsend to UK public finances. In 2008-09 it contributed £13bn in taxes, and helped contain the trade deficit. The UK's 2008 trade deficit was £44bn, but it would have been £84bn in the red without domestic oil and gas production.

Offshore oil production goes back to the start of the twentieth century in many parts of the world: off the beaches of California, on lakes between the US and Canada and between Texas and Louisiana, and off Azerbaijan in the Caspian (which in 1900 was providing half the world's petroleum).

Later on, offshore production developed in the Gulf of Mexico and on Venezuela's Lake Maracaibo. Given the innate difficulties of offshore operations, it is not surprising that there have been some major tragedies with heavy loss of life in the offshore industry. The worst incidents in the North Sea were the capsizing of the *Alexander Kielland* in the Norwegian sector in 1980 with the loss of 123 people, and the death of 167 people in the Piper Alpha blaze of 1988 in the UK sector.

Depth of water is not always the problem. Take Kashagan, a new oil field under development in Kazakhstan's part of the north Caspian. It is one of the biggest finds in recent years, and during its peak years, is eventually expected to produce more than 1m b/d.. The consortium of oil majors developing Kashagan had already spent $12bn on it by 2008, but the eventual total cost of the project is now expected to reach a staggering $136bn.

Part of the technical difficulty of Kashagan is that its oil is under high pressure in a reservoir that is more than four thousand metres below the seabed and that it contains a lot of toxic hydrogen sulphide. But the sea at this point in the Caspian is ludicrously shallow, less than ten metres deep, so that the drilling rigs sit on the seabed. Indeed the shallowness of the

The view from on top of an 80,000 cubic metre oil tank (some 500,000 barrels) at the Bolashak processing plant near Kashagan offshore oil field.

Workers prepare pipes on the *Deepwater Pathfinder*, an oil drillship that sits twelve miles off the US coast near Galveston. It was designed to operate in water depths of up to ten thousand feet.

water is part of Kashagan's problem. Winds from the Kazakh steppes have a powerful effect on the water level – the same as if you blow on a saucer of water – and in the harsh winter, wind-driven ice piles up against the rigs.

Normally, however, the bigger offshore oil reserves and problems are to be found in deeper water. According to the International Energy Agency, estimates of recoverable conventional oil lying in deepwater (depths of 400–1500 metres) or ultra-deepwater (deeper than 1500 metres) range from 160bn barrels to 300bn barrels. Most of these reserves are in the waters of Brazil, Angola, Nigeria and the US. And progress in accessing these reserves has been surprisingly fast. Deepwater oil production went from less than 200,000 b/d in 1995 to more than 5m b/d by 2007. Ultra-deepwater production only began in 2004 and is expected to reach 200,000 b/d by 2010. Chevron in particular has been breaking records in the Gulf of Mexico.

Most of the deepwater investment drilling was in the golden triangle of North America, Brazil and West Africa. Recent discoveries in the

Santos Basin off the shore of Brazil have been particularly promising. In September 2009, BP announced that it had found oil in its Tiber well in the Gulf of Mexico – one of the deepest holes ever drilled. This involved drilling down from the Tiber well itself sitting on the seabed 1259 metres below the ocean surface a further 8426 metres below the sea bed to find oil. This oil is almost as far below sea level as Mount Everest is above it.

But in April 2010 the same rig that drilled the Tiber well – the *Deepwater Horizon*, under contract to BP – sank following an undersea explosion that killed eleven crew, and released millions of gallons of oil into the Gulf of Mexico. With the well still leaking five weeks later, the Obama administration called the BP spill the worst environmental disaster in US history.

Gas

The greenest fossil

Gas has become the vital supporting element in our fossil-fuelled energy system. It is used residentially for cooking and heating. It meets 21 percent of overall energy demand, and is second only to coal in generation of electricity – gas provides a fifth of the world's power. To a small extent, as compressed gas, it is a transport fuel. It also has non-energy uses. Like oil, it is a feedstock for petrochemicals, and for agriculture it provides the ammonia used in fertilizer.

The basics

As a lighter version of oil, gas has a near-identical geological provenance (see p.47). Gas is found alongside oil (when it is known as associated gas), by itself in distinct gas fields, or sometimes seeping out of coal seams. It is frequently burned off or flared – oil companies use spouts called gas flares to discharge unexpected gas – particularly when gas comes up with oil, but not in a large enough quantity to justify a separate pipeline for it.

But in some countries, notoriously Russia and Nigeria, gas flaring takes place routinely and on a large scale. This wasteful practice is declining, but around 150 billion cubic metres of gas, or five percent of world production, are still flared off every year. This is damaging for the environment, though it's not as damaging as leaks of the uncombusted gas would be because methane, the main component of gas, is twenty times more powerful a greenhouse gas than carbon dioxide.

The latecomer

Of the three main fossil fuels, natural gas is the latecomer. Natural gas was known about in ancient times – agnostic scientists have conjectured that the vision of the burning bush referred to in the Bible may have been an incidence of a gas leak. And town gas was manufactured from coal from the early nineteenth century onwards in many cities, and used for local lighting and heating. But as a major component of the energy mix, natural gas only got going in the US from the 1920s, and in the rest of the world from the 1960s.

Will gas overtake oil?

Some people – most notably the physicist and systems analyst Cesare Marchetti – have perceived a default cycle in energy systems in which new fuels overtake old ones. Marchetti developed his Energy Substitution Model, applying statistical and behavioural analysis to the economics of energy use. He found that energy sources follow a similar trend when entering the market: that it takes forty to fifty years for an energy source to go from one percent to ten percent of market share; and that an energy source which eventually comes to occupy half of the market will take almost a century to do so, from the time it reaches one percent.

Top six gas producers in 2008 (in billions of cubic metres)

Russia	601.7
United States	582.2
Canada	175.2
Iran	116.3
Norway	99.2
Saudi Arabia	78.1
World production	3,065.6

Source: *BP Statistical Review*

According to this theory, just as coal replaced wood as the predominant fuel, and oil replaced coal, so natural gas would eventually overtake and replace oil. Marchetti's forecasts for gas have not proved accurate. In the world's energy mix, oil remains the dominant primary fuel. However, his overall forecasts for oil are not far off: oil is likely to follow a downward trend not far from that illustrated by his model.

Before there was any organized use or market for gas, drillers used to curse their luck if they found gas instead of oil, or more gas than oil. More likely than not, they would have to flare it off at no benefit to themselves and a cost to the environment. A joke from these earlier days had a geologist reporting back on the drilling of an exploratory well: "The bad news is that we didn't find oil. The good news is that we didn't find gas either."

Nowadays drillers are usually very happy to strike gas. Gas may outlast oil (though not coal). And, in contrast to oil, new gas finds still exceed production (as shown by the two tables of top gas consumers and producers here), though latterly by smaller margins than in the great decades of discovery in the 1960s and 1970s. As with oil, really big finds are fewer these days, and the smaller the field the quicker the rate of decline. In 2009, the International Energy Agency estimated that by 2030 the world would lose half its current gas production – the equivalent of two Russias – simply due to the pace of decline.

Top six gas consumers in 2008 (in billions of cubic metres)

United States	657.2
Russia	420.2
Iran	117.6
United Kingdom	93.9
Japan	93.7
China	80.7
World consumption*	3,018.7

*Production is slightly larger than consumption due to stock changes at storage and liquefaction plants.

Source: BP Statistical Review

Peak gas?

Despite these declines, we seem to be far from approaching peak gas. The global recession of 2008–09 temporarily weakened gas demand and prices. which were also weakened by the increase in US output of unconventional gas (see box). In addition, there is always the possibility of further reducing gas flaring – cutting down on gas wastage.

The best guess from the IEA is that we have only produced 13 percent of total conventional gas reserves, compared to at least 33 percent of conventional oil. Total remaining conventional gas reserves were reckoned to be around 405 trillion cubic metres at the end of 2008. The portion of these reserves that can be extracted is also higher than for oil, with minimum recovery rates of thirty percent and maximum rates approaching one hundred percent.

Unconventional gas

Conventional gas reserve estimates are dwarfed by those for unconventional gas – around nine hundred trillion cubic metres. Unconventional refers not to the chemical make-up of the gas, which remains basically methane (CH4), but the use of unconventional means to extract it. This involves special drilling and stimulation techniques to release the gas from the formations in which it is found.

▶ **Tight gas sands** These are like ordinary gas reservoirs of gas, but are less permeable and therefore harder for the gas to flow through. North America has led development of tight gas sands: this region has the largest share (38 percent) of tight gas sands reserves. US production of tight gas has steadily increased since the early 1990s.

▶ **Coalbed methane** This is natural gas, usually obtained from coal deposits that are too deep or too poor quality to be mined. Again, the US has led development of coalbed methane (around 10 percent of gas production), but in Canada, Australia and India coalbed methane gas also ranges between five to seven percent of overall production.

▶ **Shale gas** Large amounts of gas are sometimes contained in layers of impermeable shale rock. Recent advances in hydraulic fracturing, which involves creating fractures in the shale layers and then propping them open with a permeable sand, have made possible large-scale extraction of shale gas, especially in the US. But even with fancy new techniques, the share of the shale gas in place that can be extracted is much lower (8–30 percent) than for conventional gas fields (60–80 percent). A repeat of the US success with unconventional gas is doubtful, particularly bearing in mind that the more crowded communities in Europe are far less likely to accept the intrusion of heavy, noisy equipment, large demands for water and waste-water treatments – all of which are tolerated in Texas and the lower Mississippi valley – to extract shale gas.

▶ **Methane or gas hydrates** This is one of the largest – and definitely strangest – forms of unconventional gas. Methane, when combined with water at low temperatures and high pressure, creates a white crystal-like substance called methane hydrate, huge amounts of which underlie the oceans and polar permafrost. These lattice-type structures look remarkably like ice, but burn if they meet a lit match. No body knows how to extract them safely, but certain energy-resource-poor countries, such as Japan and India, are very interested in their extraction. There has been no intentional commercial production of gas hydrates. But in Alaska and Siberia some wells have been drilled through gas hydrates to reach conventional gas reservoirs, and as production lowers the pressure in these reservoirs, it is thought some of the gas hydrates have dissolved into the reservoirs below and thus been pumped out. Of course, methane is an extremely potent GHG and the release of methane hydrates due to melting Arctic permafrost could be one of the most dangerous and exponentially damaging side-effects of global warming.

Natural gas discoveries and production, 1960–2006

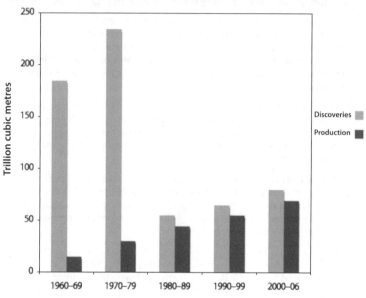

Source: International Energy Agency

Gas and the environment

The future of the extraction and supply of and demand for natural gas will depend on a number of factors besides ease of access – the environment, for one. The main component of natural gas, methane, is a more potent greenhouse gas, with greater global warming potential, than carbon dioxide. But natural gas, or methane (CH_4), has fewer carbon atoms than oil or coal. So when burned as a fuel, natural gas emits 30 percent less CO_2 than oil and 45 percent less CO_2 than coal. It is therefore the most climate-friendly of the fossil fuels.

Efficiency

Gas can be used with high efficiency in power generation with combined cycle gas turbines (CCGTs). A gas turbine works via a mixture of combusted gas and air, which expands and spins rotor blades to generate electricity.

The other output of gas turbines has always been heat, which was a wasted by-product. However, in CCGT technology, this heat isn't lost but

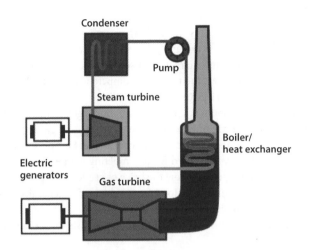

This simple diagram shows how a gas turbine generator generates electricity. The waste heat is used to make steam to generate additional electricity via a steam turbine.

used to make steam, which can turn a steam turbine to make additional electricity.

Transport

Gas also has some important drawbacks, one of them being the relative difficulty of its transporting it. At the point of actual electricity generation, gas-fired power plants are cheaper and quicker to construct than coal-fired power plants, and much cheaper and quicker to build than nuclear reactors. But transporting a gas is more complex and expensive than transporting a solid such as coal, or a liquid like oil.

Granted, a pipeline for gas may be no more expensive, in absolute terms, than a pipeline for oil. But the pipeline for gas is more expensive relative to the lower energy density of the gas, compared to oil's high energy potential. And shipping gas is more expensive in absolute terms than shipping oil or coal, because the gas has to be cooled into a liquid that is 1/600th of its gaseous volume before eventually being re-gasified at its destination. This is why gas deposits are described as "stranded" more often than oil or coal deposits: it is not that gas is necessarily found in more remote places than oil or gas: it is just that getting gas to its destination is more of a problem than for oil and coal.

Price

Another reason why gas is not making many inroads into oil's share in the overall energy mix is that gas prices generally tend to rise or fall along with oil prices. In some markets and sectors, such as power generation in developing countries, the two fuels compete with each other, and this seems to keep their prices very roughly in line.

In industrialized countries, the two fuels no longer compete much with each other (oil being mainly confined to transport, and gas to power). But in some of these markets, especially in continental Europe, gas and oil are almost contractually prevented from competing with each other by contracts in which gas prices follow the price of crude or oil products with a six- or nine-month time lag. However, gas demand is generally more sensitive to price than oil, because substitutes exist for gas in its main use for power generation, while no real alternative yet exists for oil in transport.

Regional differences and trade

If unconventional gas resources are taken into account, gas resources are fairly widely spread around the world. But some of these reserves, such as methane hydrates, may be unexploitable, and there are considerable disparities between those that can be developed.

Three countries – Russia, Iran and Qatar – hold 56 percent of conventional reserves. Russia has the largest single share (25 percent), but Iran and Qatar together have 30 percent of world reserves, and a large bubble of this 30 percent exists in one single field. Stretching across the Gulf, the Qatari portion is called the North Dome, and the Iranian portion South Pars. Yet, despite its 40 percent share of reserves, the Middle East accounts for only 11 percent of production, while at the opposite extreme is North America, which with only 4.5 percent of conventional reserves is responsible for 26 percent of world gas output.

To some extent, consumption patterns will even out in the years ahead. So the biggest absolute future increase in gas use is expected to come from Middle Eastern countries using more of their own gas at home. But they will also export more. Overall, the gas trade between the major regions of the world is forecast to nearly double in the next two decades.

Global gas through LNG?

In one sense, liquefied natural gas (LNG) is the solution to the problem of transporting a gaseous form of energy. Freezing natural gas to minus 163

A portrait of the late founder of the Islamic Revolution, Ayatollah Ruhollah Khomeini, hangs over an office in the South Pars Special Economic Energy Zone, in Asalouyeh, south Iran, situated on the coast near the large offshore South Pars gas field.

degrees centigrade, and thereby liquefying it, also has the effect of shrinking it to 1/600th of its gaseous volume.

But overcoming the transport problem for natural gas out of reach of a pipeline is a very expensive business. It requires big investments in four links of a chain:

▶ **The usual set-up costs** associated with any gas field development.

▶ **The liquefaction process** which involves passing gas through something that is called a "train" but only resembles a railway train in being a long series of chillers and liquefiers of the gas. Units of LNG plants are known as "trains".

▶ **Transporting it by sea** in cryogenic or specially insulated tankers.

▶ **Warming the natural gas** in a re-gasification terminal.

The cost of this LNG chain can be anywhere between $10bn and $20bn for an project generating 8 million tonnes per year, of which the liquefaction process accounts for about half the expense. It is therefore not surprising that the LNG business is still dominated by the oil majors plus one or two gas specialists such as BG Group of the UK (which was once the upstream division of the old British Gas). However, there is one gas exporting country, Qatar, which has built and owns a whole LNG chain including tankers and a re-gasification terminal in the UK.

The sphere of a liquefied natural gas tank at a plant near Charleston, South Carolina, destined to be fitted into a tanker ship.

The rich coastal states of northeast Asia (Japan, South Korea and Taiwan) have traditionally been the major markets for LNG, but there are now 17 countries that import LNG and 15 that export it. Europe, especially Spain, has been a growing market, and in the mid-2000s there was an expectation that the US would start to draw in major amounts of LNG and many re-gasification terminals were built. But in the last few years, there has been a surge in production of US unconventional and tight gas, which has made a lot of American LNG surplus to current domestic requirements.

Unfortunately, LNG is not an industry that can respond very flexibly to changes in demand, because it has traditionally worked on long-term contracts, sometimes as long as twenty years, between buyer and seller. This kind of point-to-point contract, in which every LNG cargo has a pre-ordained destination, gives developers and their financiers security of demand and security of supply which has been very important to the cash-rich but energy-poor countries of northeast Asia to the customer.

The gas OPEC – a mirage?

Ever since a group calling themselves the Gas Exporting Countries Forum (GECF) started meeting informally in 2001, gas importing countries have been spooked by the fear of a "gas OPEC" being formed as a cartel to make them pay more for their gas. But while the GECF may not always be a paper tiger, it hasn't really growled yet.

In 2008 members of the GECF, including the big three – the "gas troika" of Russia, Qatar and Iran – put their association on a more formal footing. They set it up as an international organisation with a headquarters in Doha, Qatar, with pledges of cooperation on all aspects of the gas industry, although nothing explicit was said about prices. It is indeed hard to see how the gas producers could replicate OPEC, which influences prices indirectly by regulating production. The GECF countries would find that very hard to do, because gas producers' ability to vary the volume or prices of gas exports is limited by long-term contracts. Direct dictation of prices would be even harder in gas than in oil.

OPEC tried to set an administered price for oil, but gave up the attempt in the mid-1980s. But at least with oil there is something like a global balance of supply and demand. In gas, this balance is always regional, with different demand and supply conditions in North America, Europe and the Asia-Pacific region, and therefore different prices. The GECF countries could usefully coordinate their projects to build LNG, the one element in the gas trade that links all these regions. But even here, common interests are not obvious. Even within the gas troika, there are big discrepancies in resources and methodology: Qatar exports all of its gas by LNG, while Russia is only just beginning to try a very small amount of LNG and Iran, bizarrely, is still a net gas importer.

But some companies, such as Britain's British Gas (which has no oil to distract it), have become big enough in LNG that they can offer security of supply to customers out of their portfolio of LNG contracts. In this case, a customer does not have to know where their LNG is coming from – they just need to know they have a firm contract with BG and that BG always has enough LNG cargoes afloat to fulfill demand.

In parallel with this, a sort of spot market (a market in which commodities are traded and delivered immediately, rather than at a point in the future) has developed for uncontracted LNG gas, with cargoes being drawn to whichever region – North America, Europe or Asia – will pay the highest price at any one time for spot cargoes.

It gives rise to the question whether LNG trade will one day equalize prices between regions to produce a global gas price, as exists with oil. It's possible. Certainly LNG trade tends to equalize LNG prices across the various regions of the world, but whether it can have the same effect on general gas prices is doubtful. LNG amounts to only about ten percent of the overall gas market, which is dominated by pipelines and marked by regional price differences; it will probably stay too small relative to the whole market to influence it. The tail would have to grow a lot bigger before it could wag the dog.

Coal

Deep down and dirty

Coal is the very widespread residue of ancient plant remains, most of which have been transformed over time into first peat, then lignite and finally high-quality anthracite. It is the largest energy source for generating electricity, accounting for around 47 percent of the world's power.

The basics

Coal is extracted by deep mining (tunnelling underground) or strip mining (stripping off surface soil and mining the exposed coal seams with huge bucket and shovel machines). Apart from making electricity and heat, a variant of coal – coking coal – is vital to making steel. It used to be the main source of synthetic gas, known as town gas because it was stored in large tanks dotted around towns and cities. Town gas has now been totally replaced by natural gas but the cylindrical structures that once surrounded the gas tanks are still to be seen in many UK cities.

The top six coal producers in 2008 (in millions of tonnes oil equivalent)

China	1,414.5
United States	596.9
Australia	219.9
India	194.3
Russia	152.8
South Africa/Indonesia	141.1 (each)
World production	3,324.9

Source: *BP Statistical Review*

Workers stack coal bricks at a coal plant in Shanxi province, China.

Top six coal consumers in 2008 (in millions of tonnes oil equivalent)

China	1,406.3
United States	565.0
India	231.4
Japan	128.7
South Africa	102.8
Russia	101.3
World consumption	3,303.7

Source: *BP Statisical Review*

Coal's alarming comeback

In climate terms, coal is the scariest of the three main fossil fuels. Of the three it is the least likely to run out, it has the fastest growing demand and does the most damage to the atmosphere. Coal has indeed become the "new nuclear", in the sense that across North America and Europe at least, proposals to build new coal plants are now more likely to stir controversy and protest than plans for new nuclear reactors.

As a primary energy source, after World War II coal lost to oil the number one spot that it had held since the 1700s. But it is now staging a comeback. Between 2000 and 2006, coal consumption increased by 4.9 percent per year, due mainly to extraordinary growth in China. The IEA forecasts that over the next 20 years or so (2006–30), growth in coal demand will average 2 percent a year, compared to 1 percent for oil, 1.8 percent for gas and 1.6 percent for all types of energy. It is estimated that almost all (97 percent) of this growth in demand will come, from developing countries, and from two countries in particular – China and India.

China – king of coal

China's output and consumption of coal has become truly staggering. The country now produces 2.8 billion tonnes of coal a year, or 46 percent of the world's total. Since 1997 China's annual coal output has increased by 1.1 billion tonnes – or more than the US's entire production in 2007. This has added an extra 2.2 billion tonnes of CO_2 every year to China's greenhouse gases, now the largest of any country. The annual growth of its production has been over 200m tonnes, or nearly Russia's total annual output.

China's coal production has a huge impact on world coal prices. Any change in its appetite for foreign coal has a big effect on the world coal market, because the internal market for coal in China is now three times the size of the total world seaborne coal trade. Nearly eighty percent of China's total electricity generation is coal-fired (a ratio only exceeded by Australia, South Africa and Poland). Coal's predominance in power generation is expected to continue, and its use is expected to grow at 4.9 percent a year up to 2030.

Supercritical coal

The IEA reports that China has made "considerable progress in the implementation of state-of-the-art coal-fired technologies". About half of all the new coal plants China is building have so-called "supercritical" technology, using very high temperatures and pressures to generate electricity. This boosts efficiency by getting more energy out of a given amount of coal. The reason why China is now beginning to overtake countries such as the US or Britain in the portion of its coal plants with supercritical technology is that its high rate of plant renewal or building gives it the opportunity to introduce new technology. For the same reason, half of all its coal plants also now have flue gas desulphurization (FGD) equipment: as its name suggests, this takes the sulphur dioxide out of the flue gases which escape up the smoke stack. Sulphur dioxide is not a greenhouse gas, but a cause of acid rain, which strips leaves off trees.

But all this new technology only reduces the relative impact of China's surge in coal use. By doubling the absolute level of its coal consumption over the past decade, China has overtaken the US to be the largest single pumper of carbon into the atmosphere.

Moreover, although China's rulers in Beijing are imposing more modern technology on new or larger power plants, they are having difficulty getting their orders for the closure of smaller, older, dirtier coal mines and power plants carried out across their huge country.

Longwall mining machines consist of multiple coal shearers mounted on a series of self-advancing hydraulic ceiling supports. They extract "panels" – rectangular blocks of coal as wide as the mining machinery and as long as 12,000 feet (3650 metres).

Coal, coal everywhere

Certainly there will not be any shortage of coal in the near future. It is the most abundant of the fossil fuels. Estimates of total potential resources some of them uncommercial and of the narrow category of proven reserves (considered commercial) have gone up in recent years. According to figures established in 2009 by the German Federal Institute for Geosciences and Natural Resources, economically recoverable reserves of hard coal are 729 billion tonnes and – on top of that – total potential resources are around 16,000 billion tonnes.

Likewise, reserves of brown coal or lignite amount to 269 billion tonnes and, in addition, total lignite resources amount to over 4,000 billion tonnes. Reserves and resources on this scale would last the world at least 150 years at the current (2009) rate of production, even assuming China maintains its present enormous output. In fact, burning even a small fraction more of these reserves will put the climate in severe danger. As the IEA says in its 2009 *World Energy Outlook*, "limits to the use of coal come not from any lack of reserves, but from logistical factors and – above all – from the environmental effects of its use".

Why is coal staging a comeback? Part of the answer lies in problems with alternatives: in recent decades nuclear operations have seemed less safe and gas supply less reliable. But the fundamental reason why coal is regaining a share in the world energy mix is because it is cheaper to produce as a fuel than oil and gas. It is also as easily transportable as oil and more so than gas, and is cheaper to use as a fuel than uranium, the fuel source of nuclear power plants, because it does not need a complicated reactor.

Today's coal mining is increasingly sophisticated – with longwall cutting machines for mining underground seams and huge cranes for strip mining, as well as some legacy Dickensian shovel-and-pick operations. The capital costs of digging coal remain lower than the expense of extracting oil and gas. World coal prices have kept pace with trends in oil and gas prices sometimes because of contractual linkages and sometimes because some power plants can switch back and forth between coal and gas. But the peaks in coal prices have had less to do with capital costs and supply issues and more to do with coal's popularity.

Coal also remains in use because of issues of national (and international) energy security. The geographical concentration of coal is greater than for oil or gas. The countries that are the top ten coal reserve-holders also hold 92 percent of the solid black stuff; the top ten in gas hold 76 percent of gas reserves, while the top ten in oil account for only 54 percent of oil reserves. The top four coal reserve-holders are in descending order the US, Russia, China and India, and they are all big energy users. But such is the abundance of coal that the remaining 8 percent of coal reserves (outside the top 10 countries) spreads a lot of energy security among other countries.

Moreover, societies as well as states have proved very loyal to coal mining. It is surprising that such a dirty and dangerous way of earning a living should attract loyalty. But whole communities have grown up around coal mining, the oldest branch of the energy business, and these communities

CCS – how it works

CCS is a three-step process. First, CO_2 is captured from the power plant. Then it is compressed and transported to a storage site, which could be a depleted oil or gas field, mined-out coal seam or deep saline formation (porous rocks which but with briny salt water to seal the CO_2 in salt is a very good sealant).

Natural gas has been stored in salt caverns, and the US government stores the one billion barrels of oil of its Strategic Petroleum Reserve in salt caverns in Texas and Louisiana. Finally, it is injected into the storage site.

There are three ways of capturing the CO2:

▶ **Pre-combustion capture** This process does not burn the coal, but chemically separates it into CO_2 (for piping into storage) and hydrogen which is then burned to generate electricity in a combined cycle gas turbine. This could also be useful if hydrogen were to become a feedstock for fuel cells in cars and houses, as well as for burning in power stations. However, it is not a process that can be retrofitted to existing power plants built for burning coal.

▶ **Post-combustion capture** This involves removing the CO_2 and absorbing it into a chemical solvent as it comes out with the flue gases from coal power-stations or industrial plants. It's a process that can be retrofitted. Retrofitting will be enormously important in curbing emissions from all those plants going into service now, before CCS becomes widespread. However, it requires that plants being built today should be capture-ready – that is, they should have the space alongside and inside the plant for the CCS equipment to be added at a later date.

▶ **Oxy-combustion process** This entails burning the fossil fuel which could be natural gas or coal in a mixture that includes pure oxygen, which turns carbon monoxide into carbon dioxide; this can then be separated out for storage.

naturally resist becoming disenfranchized, as was evident in the UK during the miners' strike of 1984.

The push for cleaner coal

In a way it's rather shocking that it took so long for the world to become concerned about making coal less dirty and dangerous. Deep underground mining kills several thousand people around the world each year; surface strip mining requires enormous amount of remedial work to prevent it permanently scarring the landscape; and coal-fired power plants throw out sulphur dioxide and nitrogen oxide emissions that require special control if they are not to contribute to human disease, smog and tree defoliation.

Some of these bad consequences of coal use are what economists call "externalities" – meaning a spillover from an activity to those not directly involved in that activity. In this case, these would be nitrogen oxide from coal (causing smog) or sulphur dioxide from coal (causing acid rain). But it took the "globality" of the carbon dioxide coming from coal burning, causing climate change, to stir global concern.

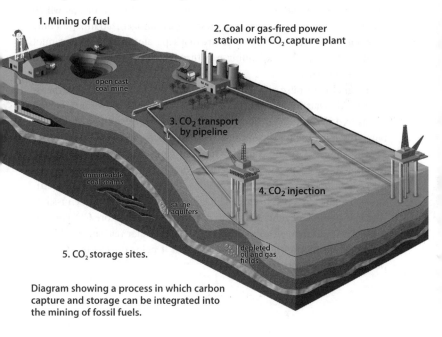

1. Mining of fuel

2. Coal or gas-fired power station with CO_2 capture plant

open cast coal mine

3. CO_2 transport by pipeline

unmineable coal seams

4. CO_2 injection

saline aquifers

5. CO_2 storage sites.

depleted oil and gas fields

Diagram showing a process in which carbon capture and storage can be integrated into the mining of fossil fuels.

The big idea to clean up coal is the capture and storage underground of carbon dioxide, a process known as CCS. This technology is especially applicable to coal-fired power plants – much more so than gas or oil-fired plants. The considerable extra expense of CCS technology is easier to justify in the case of coal-fired plants, because they emit a high proportion of CO_2 in their flue gases. There could also be commercial spin-offs. Surplus CO_2 could be used to encourage cultivation in closed greenhouses where plants re-absorb the CO_2, to enhance oil recovery – one potential negative spin-off that environmentalists are rightly concerned about – and, depending on the CCS process, to produce hydrogen for use in vehicle fuel cells.

More efficient coal

The extra cost of CCS lies not only in the technology, but also in the power needed to run the technology, which effectively lowers the output of the generating plant. In fact, the industry has been working for some time to improve efficiency in power generation, in particular in the pulverization of coal so that it burns more rapidly and efficiently and using "supercritical" and "ultra-supercritical" coal combustion technology. This technology heats and pressurizes water above its "critical point" (at which water is still separable into steam), produces a homogenized mix (not only very hot, but under such high pressure that separate steam bubbles cannot form) and passes this expanding mix through to rotate a turbine and so generate electricity.

The goal is to raise efficiency ratios – the output of electricity (measured in kilowatt hours) as a proportion of the input of primary energy (measured in joules). The aim is to lift these ratios into the forty to fifty percent range (and maybe one day above fifty percent), up from the thirty percent standard of, for example, the older Chinese generating plants, or the mid-thirties percentage range typefying much UK and US electricity generation. Every percentage-point improvement in generating efficiency equates to a two to three percentage-point reduction in emissions (because inefficient generators emit disproportionately more CO_2 per unit of energy input than efficient ones). But the particular relevance of higher energy conversion efficiency to CCS is that it helps offset the CCS cost penalty in lost electricity output.

Coal into gas

Coal can also be turned into a gas – above ground in one process and below in another. The more widely developed above-ground process is called an Integrated Gasification Combined Cycle unit. This turns coal, as its name suggests, into a synthetic gas.

This synthetic gas can then be separated into CO_2 for CCS storage or for industrial use in chemicals and plastics and hydrogen (for use in gas turbines). That is the "gasification" part of the IGCC process. The "combined cycle" part of it is something that has created a minor revolution in electricity generation efficiency. This takes the exhaust gas from the gas turbine (which functions a bit like a jet engine) and passes it through a heat-recovery steam boiler that drives a second turbine, whose electricity output is combined with that of the gas turbine. When the original fuel is gas, not coal, this technology is known as a Combined Cycle Gas Turbine (CCGT).

The technique for producing synthetic gas or syngas from coal is called underground coal gasification. The aim here is to burn coal underground, especially in seams too deep or thin to be easily mineable, to produce synthetic gas, separable into hydrogen for a CCGT and CO_2 for CCS.

Coal into oil

A much older technology, Coal-to-Liquids (CTL), can create synthetic oil out of solid coal. This was originally exploited by states that had no access to conventional oil. It was especially useful to Germany after it lost access to oilfields in the closing years of World War II, and to apartheid South Africa after the 1950s after it became subject to international trade sanctions.

But it is hard to see how CTL could be adapted to carbon capture and storage, and even if it could not be, it produces more emissions in the manufacture and use of the synthetic oil than burning the coal in the usual way.

Carbon capture and storage: wish fulfilment or wishful thinking?

The great advantage of carbon capture and storage (CCS) technologies, if they work, would be to prolong the fossil-fuel age, with all its convenience. As the Stern Review noted in 2006, "CCS technologies have the significant advantage that their large-scale deployment could reconcile the continued use of fossil fuels over the medium to long term with the need for deep

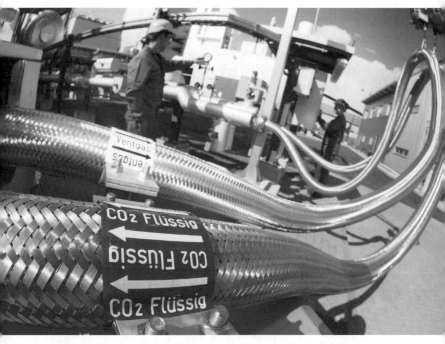

Two employees at the Schwarze Pumpe ("Black Pump") power station near Berlin, a pilot CCS project operational since 2009. The pipes carry liquid CO_2 which can then by sequestered via injection into saline aquifers, deep within the ground.

cuts in emissions". It would also, the review said, "effectively reduce emissions from the flood of new coal-fired power stations planned over the next decades, especially in India and China". But the very convenience of CCS working should put people on their guard against wishful thinking.

CCS technologies have been tried before. Capture technologies have been used in the making of chemicals and fertilizers, and to take the CO_2 out of natural gas. There has also been some limited CCS useage in the oil and gas sector. CO_2 is being re-injected into natural gas fields in the North Sea and onshore in Algeria. It is also being piped from a coal gasification plant in North Dakota in the US to a Saskatchewan oilfield in Canada to increase oil recovery there.

The novel and daunting aspect of CCS will be to combine capture, transport and storage across the entire power sector. One UK expert likened it, in the British context, to "carrying out the whole desulphurization of all UK power plants, plus development of North Sea oil and gas network, all over again". In the US the Department of Energy has mapped out possible CCS storage sites across the country, and concluded that the majority of major carbon emitters, such as coal power plants or heavy industry, have suitable storage within fifty miles.

Nevertheless, the US, which already has a network of some 3500 miles of CO_2 pipelines (mainly for oil recovery), would have to multiply this network a hundred times if CCS were fully implemented. The eventual CO_2 pipe network would be "the equivalent of today's entire US natural gas network", according to one US expert.

The world has never before had to deal with the environmental problems of storing CO_2 underground. Some critics of CCS have cited the Lake Nyos disaster of 1986. Lake Nyos is a lake in the flank of a volcano in the African state of Cameroon, under which a large body of CO_2 built up. Suddenly, on 21 August 1986, the lake emitted a cloud of CO_2 which effectively suffocated some 1700 people and killed several thousand livestock. The tragedy was probably very specific to this volcanic site. Nonetheless, local objections are increasing, particularly in Germany, from people worried about having CO_2 buried underneath them, giving rise to a new variant of NIMBYism: NUMBYism (not *under* my backyard).

CCS is manifestly not the highway to a permanent low-carbon economy. We could not go on forever burying CO_2 underground – because we would run out of convenient and safe storage – in the way that we could go on generating electricity with the power of the wind or the sun. (And, long-term, renewables would be cheaper and more efficient.) But CCS does offer the best short- and medium-term hope for dealing with emissions from coal power stations and other big industrial sources of CO_2.

Governments in North America, Europe and China are making a start. The Obama administration set aside $2.4bn for CCS test-projects in its green energy stimulus package, and the European Union has earmarked about the same sum for CCS in immediate funding, passing legislation that would eventually give CCS operators some emissions permits they could sell. Individual EU governments are doing more.

The UK has now committed itself to funding four test projects and has also gone further than EU legislation to require that no new coal power plants can go ahead without applying CCS to a substantial part of their output; and that all coal plants will have to retrofit CCS within five years of CCS being "independently judged to be technically and economically proven".

For the future, three things are clear:

▶ **CCS will need a subsidy** to take it beyond demonstration to deployment, and to the point where the cost of CCS falls below the price of carbon permits on the US and European carbon trading systems.

▶ **CCS is an option for developed countries,** and China, which appears to accept the climate responsibility that comes with mining nearly half the world's coal. The most that other developing countries, such as India, will do – but must do – is to ensure that their new coal plants are at least capture-ready.

▶ **We need a Plan B** if CCS fails.

Nuclear power

Better the devil you know?

Nuclear power accounts for only around six percent of total primary energy demand in the world, but it generates about fifteen percent of global electricity. The main fuel for nuclear power is uranium, a mineral that is fairly widespread around the world. Some twenty countries mine it. Kazakhstan became the biggest producer in 2009, with Australia and Canada close behind. Namibia, Niger, South Africa, Russia and the US are also significant producers.

The basics

Built on the scientific discoveries about atomic physics that were made between the two world wars, which led to the atomic bomb and then the hydrogen bomb, nuclear reactors have been used to make electricity since the mid-1950s. By the end of 2009, there were 436 reactors in 30 countries, with 52 more under construction and a further 135 planned, according to the World Nuclear Association. Some 150 naval ships around the world are also powered by nuclear reactors.

Reactors in nuclear power plants work just like any thermal power plant – heating water into steam that drives a rotating turbine which generates electricity. What creates the heat, in nuclear fission reactors, is the release of energy when uranium atoms are split. There is an enormous amount of energy in the bonds that hold atoms together, and when these atoms are split, a massive amount of energy is released. The reason why uranium is used is that its atoms are easier than those of other minerals to split apart.

The top six nuclear power producers

	Number of reactors operating	Megawatts
United States	104	101,119
France	59	63,473
Japan	53	46,236
Russia	31	21,743
Germany	17	20,339
South Korea	20	17,716
World	436	372,900

Source: World Nuclear Association

A second chance?

Although new coal plants provoke direct protest action and peak oil always causes a bit of intellectual friction, no energy issue has stirred more emotion and controversy than nuclear power over the years. This is hardly surprising, because there are strong arguments, strongly held, for and against it.

We have lived with nuclear electricity for more than fifty years now. Some 440 reactors in 30 countries generate about 15 percent of the world's power, although in a few, mostly richer, countries the proportion of nuclear generation is higher. In France the nuclear share in electricity generation is 80 percent, and France, the US and Japan account for 57 percent of world nuclear generating capacity. After the 1970s oil-price shocks, France and Japan launched major reactor-building programmes for reasons of energy security, well before scientists began to worry about climate change.

According to the World Nuclear Association, there were, in 2009, some 380 new reactors planned or proposed around the world. Although the rate of reactor building has slowed in recent years, nuclear power still makes a powerful contribution to low-carbon energy generation, because of its scale. Nuclear power generates one third of the European Union's electricity, and accounts for two thirds of its CO_2-free electricity. The Nuclear Energy Agency, an offshoot of the OECD, has estimated that if nuclear power replaced coal, it could be "saving" between four and twelve gigatonnes (billions of tonnes) of CO_2 a year by 2050, depending on how nuclear power grows in the coming years.

For nuclear power	Against nuclear power
It's an extraordinary energy source: 1 tonne of uranium is equivalent to 16,000 tonnes of oil.	The reaction process can spin out of control with disastrous results, as accidents such as Three Mile Island and Chernobyl have shown.
Compared to the thousands killed every year in coal mining, a relatively tiny number of people have been killed in nuclear incidents.	Always the potential, however slight, for catastrophic accidents. and the risk of bomb-making technology spreading.
Nuclear energy produces no carbon emissions.	There is a carbon footprint attached to the building of reactors, and an overall environmental cost to uranium mining.
The volume of nuclear waste is small, and therefore manageable.	The management of nuclear waste and decommissioning of reactors is lengthy and expensive.
If other fuels had their external costs (such as pollution and health) included in their price, nuclear would be competitive	Nuclear power was massively subsidized by governments in the past, and is fundamentally uneconomic.

Meltdowns and markets

But the nuclear industry took several hits in the 1980s and 1990s. The Three Mile Island accident in 1979, in which a nuclear generating station in Pennsylvania suffered a partial core meltdown, brought US reactor construction to a halt and was a significant factor in Sweden's vote the following year to phase out their reactors by 2010. The far more serious accident in Chernobyl in 1986 turned many more European nations off nuclear power. In a 1987 referendum, Italians voted to end their nuclear programme definitively, while Germany, Spain and Belgium subsequently decided to run their nuclear programmes down.

Energy liberalization and privatization in Europe and the US also had an impact on nuclear power. This tended to favour quicker investments with lower capital costs, such as gas-fired power plants, and to penalise more expensive and inflexible investments such as new nuclear reactors. Most existing nuclear operators continued to make money out of existing reactors, whose costs had been all paid off. (In 2003, however, seven years

History of the global nuclear power industry

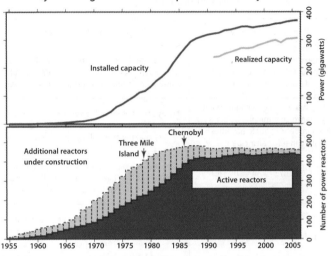

Source: International Atomic Energy Agency/Robert A. Rohde

after it had been privatized, British Energy had to be rescued by the UK government, which in 2008 sold BE to Electricité de France.) But, through most of the 1990s and 2000s, utilities shied away from investing in new reactors in the US and Europe. Only in Asia did orders for nuclear reactors continue.

Domestic and international concern about the safety of nuclear reactors tends to fluctuate alongside major accidents. But lessons have been learnt from these incidents. Since the Three Mile Island accident, the nuclear industry in the US has consolidated into bigger and more technically competent companies. Since the Chernobyl accident, similar reactors without outer containment shells have been closed down. The safety record of reactor operators has generally improved.

However, public concern about the disposal of nuclear fuel waste remains a constant, because the problem remains a constant. Many countries are wrestling with the problems of permanent repositories for their nuclear waste. But only one country, Finland, has taken a final decision on a permanent burial site for its waste. The Onkalo tunnel, dug in low forest on the western coast of Finland is the entrance to a spiralling track that reaches down through solid rock to a depth of five hundred metres. A system of multiple barriers – most of them engineered, one of them being natural solid crystalline rock – protects a series of horizontal shafts right

These copper "coffins" will be the final resting place for Finland's uranium rods. They are thick enough to resist corrosion for many hundred thousand years.

at the bottom. The actual nuclear fuel rods will be placed in corrosion-resistant copper canisters, of a thickness designed to secure the waste for hundreds of thousands of years.

The US is exploring different security technologies at its Yucca Mountain site. Meanwhile, in Finland as everywhere else, nuclear waste is still piling up in temporary storage beside the reactors that produced it.

"Too cheap to meter!"

The second most troubling question about nuclear power is its economics. Nuclear power is the only energy technology that seems to have become more, not less, expensive over time. The energy company EDF estimates its latest reactors will cost €4–5bn. How different this is from the 1950s when, in the nuclear industry's first flush of enthusiasm, it was thought that nuclear power would be "too cheap to meter" and that it would inevitably undercut other forms of electricity generation.

In reactor building, France achieved some economies of scale by going for standardized designs in its crash reactor-building programme in the late 1970s and 1980s. But even the French are still facing a big cost over-run on the first-of-a-kind Generation 3 reactor that they are building in

Finland. The UK nuclear programme has been plagued by a piecemeal approach, building two types of UK-only reactors (Magnox and Advanced Gas Cooled Reactors or AGRs) but with different and costly modifications to every one of them, while the US nuclear industry suffered from being too fragmented into small operating companies.

The economics of the nuclear industry became less competitive with energy-market liberalization. The notion, especially in the US and the UK in the 1980s and 1990s, that a country's energy mix should be established by free competition between energy sources in the marketplace, appeared to reveal the fundamental inability of nuclear power to survive without government subsidy. The nuclear industry, which has to lay out several billions of pounds, dollars or euros before a reactor can start earning its keep, has found the market uncertainties created by energy-market liberalization hard to cope with.

Liberalization has had to cede ground in the face of the climate-change challenge, which has itself given a whole new lease of life to – or at least one new argument for – nuclear power. For nuclear power is the only major source of carbon-free electricity with a proven record of power generation on the scale required.

Even staunch opponents of nuclear concede that nuclear power generation is carbon-free, although they point out that the construction of reactors inevitably involves substantial emissions – and that uranium extraction itself has an environmental footprint.

The nuclear renaissance

Rebirth in the US

After some false dawns, it looks as though there is something of a nuclear revival underway. After a 28-year period (1979–2007) in which there was not a single US licence application for a new reactor, requests for 27 new reactors have been filed since 2007 with the US Nuclear Regulatory Commission. Approval of the first of these licence requests is expected some time between 2011 and 2012. And the industry, gearing up for new construction, has hired some eighteen thousand extra people.

The only hiccup for the US nuclear industry is that the new president, Barack Obama, has come out against the long-term project to make Nevada's Yucca Mountain the final repository for civil nuclear waste. In doing so, Obama is honouring a 2008 campaign pledge he made to win the support of Senator Harry Reid of Nevada, the Democratic majority leader, who opposes the plan to bury high-level atomic waste at Yucca. But the US would hardly be the first country to start, or resume, building reactors without knowing where the final burial site for their high-level waste will be. Only Finland, as pointed out earlier, has made that decision. (In fact, the US is already burying its military nuclear waste at a site in New Mexico.)

Revival in the UK

The UK, which was the first country in the world to start up a commercial nuclear reactor (in 1956), has nineteen reactors providing about twenty percent of the country's electricity. But fourteen of them will have to shut down before 2020, and probably all but one by 2023. The reason is a mixture of age, inefficiency and unreliability.

Britain's remaining Magnox reactors started operation in the late 1960s and early 1970s, and its AGR reactors began working in the late 1970s and in the 1980s. Most of today's reactors in the world were originally

The dome containing the reactor at Sizewell B nuclear power station. The UK's only commercial pressurized water reactor power station was built and commissioned between 1987 and 1995.

designed for a life of up to forty years, and many have now had their operating lives extended for a further twenty years. In the US, for instance, some fifty reactors have had life extensions to sixty years. But the Magnox and AGR designs are specific to the UK, and have generally not worked well. Some Magnoxes closed earlier because they were uneconomic, and the AGRs are considered very unlikely to get life extensions.

Ironically, the only UK reactor that might, on the US precedent, get a life extension is the most recent – Sizewell B on the Suffolk coast, which started operating in 1995 and would be due to shut down forty years later, in 2035. Sizewell B is Britain's only Pressurized Water Reactor, the model for most of the world's reactors in the US and elsewhere that have been given life extensions by nuclear regulators.

At the same time, the UK has committed itself to very stringent emission cuts. Not only is the UK party to the European Union commitment to reduce emissions (from 1990 levels) by twenty percent by 2020, but by separate national legislation the UK has bound itself to achieve an eighty percent reduction (from 1990 levels) by 2050.

There is no (realistic) alternative

The UK has no chance, on its current trajectory, of meeting its longer-term emission targets while giving itself some energy security, unless it builds new reactors just to replace the older ones coming off-line. If the UK had no nuclear power today, its current emissions would be between five and twelve percent higher, depending on whether the replacement power came from gas-fired power stations or dirtier coal-fired plants. If you wanted to save even that 5 percent of extra emissions, you would have to take a third of the UK's 32 million cars off the road.

The UK has, relative to other countries, very ambitious national emission reduction goals. British political parties are likely to want to achieve this by being particularly tough on decarbonizing the electricity sector, so as to minimize cuts in aviation – the UK is after all an island with only one international rail link. To achieve an eighty percent overall emission cut by mid-century is reckoned to require an electricity sector that would be forty percent low-carbon (generated by nuclear and renewable means) by as early as 2020.

Most of this increase is already supposed to come from renewable energy, whose share of electricity generation is to rise from 5.5 percent today to 30 percent by 2020. It is hard to imagine renewable energy managing this six-fold increase in its share of the UK's power mix over the next decade. It is impossible to imagine renewables achieving not only this, but also taking over nuclear's share in the same relatively short period.

So, in line with this logic, the UK government has taken a series of decisions regarding its nuclear industry. In 2003 it rescued British Energy, the UK nuclear operator, from bankruptcy. In 2006 it announced legislation to speed up certification of new nuclear designs and to streamline the planning process for new reactors. In 2008 British Energy was sold to Electricité de France, a company with the financial and technical means that the UK company no longer had. In 2009 it announced its approval of a series of sites for new reactors to be built.

But it still remains to be seen whether swifter licensing and planning procedures will be sufficient incentive for new reactor construction. The nuclear industry points to the public subsidies now going to developers of renewable and clean-coal technologies, and asks for the same. Yet nuclear lacks the public acceptance that makes it possible for politicians to stand up and justify a public subsidy for it.

Reversal in Europe

For the same climate change and energy security reasons influencing the UK, three other European countries – Germany, Sweden, Belgium – have all, during 2009, reversed or delayed earlier decisions to phase out nuclear power. Sweden reversed a decision, taken after a 1980 referendum, to phase out the country's nuclear reactors. Belgium overturned a 2003 limitation on the country's reactors to no more than forty years of operation, and postponed the closure of three of its oldest reactors from 2015 to 2025. In Germany the centre-right coalition has delayed the phase-out, decided by the previous centre-left coalition, of the country's 21 reactors by 2021.

But none of these three countries has decided to build a new generation of reactors. For its part, Spain is sticking to its 1983 moratorium on the building of any new reactors. Only Italy, which in line with a 1987 referendum decision has entirely halted nuclear power, is planning to start from scratch again and build a new generation of nuclear plants with French help.

Improving nuclear power

New generations in nuclear fission

The fact that researchers are now developing what they call the fourth generation of nuclear reactors will come as a surprise to many people who will ask: what happened to the first three? Well, actually they are mostly still with us – apart from the first generation, the original power reactors built after World War II, which have all been withdrawn from service.

Weapon worries – a constraint on civil nuclear power

The growing concern about Iran and North Korea acquiring the bomb-making know-how and capacity for nuclear warheads is making it harder for nuclear newcomers to embark on civil nuclear power programmes. The concern is clearly justified. North Korea has boasted that it has made several bombs. Iran insists that its nuclear programme – including its efforts to master the full cycle of uranium fuel enrichment – is purely peaceful. But the suspicion is increased by the fact that the expensive business of trying to enrich uranium makes little economic sense for a country such as Iran, with just one half-built Russian reactor, unless bomb-making is an eventual aim.

On a political level, Iran and North Korea have devalued the Non-Proliferation Treaty (NPT), which came into force in 1970 to stop any further spread of nuclear weapons beyond the "big five" – the US, Russia, China, Britain and France – which already had them by that date. Both Iran and North Korea signed the NPT, in contrast to three – Israel, India and Pakistan – which never signed and decided to retain and exploit their diplomatic freedom to build nuclear weapons.

The likelihood of a world-wide increase in civil nuclear power, for climate reasons, will inevitably create concern about weapons proliferation, and nowhere more so than in areas of tension such as the Middle East. A number of Arab and Gulf states now want nuclear power in order to desalinate water and to conserve domestic gas for higher-value purposes such as raw material for petrochemicals.

The country that has advanced furthest down this road is the United Arab Emirates, led by Abu Dhabi, even though it is the UAE's main oil producer. Abu Dhabi has signed various international nuclear cooperation agreements; they contain a promise by Abu Dhabi that it will not enrich nuclear fuel or reprocess nuclear waste, but instead import enriched fuel from abroad and return the waste abroad – as the international community would like Iran to do. In November 2009, Abu Dhabi signed a contract with a South Korean-led consortium for the construction of its first reactor.

The reactors in operation today are dubbed second generation, while the reactors that will be built over the next ten to fifteen years are generally described as third generation, even though they are refinements of second generation reactors. The vast majority of second – and third – generation reactors are pressurized water reactors or boiling water reactors. These have two main drawbacks:

▶ **An inefficiency** in the requirement for a secondary loop of water to be heated (in order to produce turbine-turning steam) by the primary loop of water that passes through the reactor core.

▶ **Production of a lot of waste** from fuel that passes only once through the reactor and is then stored.

So work is underway to design a fourth-generation reactor that helps to ease one or other of these problems.

One of these designs is a supercritical water-cooled reactor (SCWR), which heats water under pressure – as we have seen in supercritical coal plants – beyond its critical point, becoming a homogenous fluid that can drive a turbine directly, without steam. Another is the high temperature gas-cooled reactor (HTGR), which heats helium, and which can then be used to drive a turbine directly. The other main family of possible fourth generation reactors are called fast reactors("fast" refers to the type of neutrons in the nuclear reaction). These fast reactors are variously termed sodium-cooled, gas-cooled and lead-cooled, depending on the type of coolant used. (It should be noted that coolant is a bit of a misnomer here, being fantastically hot: the coolant conveys the heat from the nuclear reaction.) The advantage of fast reactors is that they can burn up nuclear waste or drastically reduce it in volume and toxicity. The disadvantage is they can be used to "breed" weapons-grade plutonium.

Future fusion – fantasy or feasible?

There is an alternative to nuclear fission. It's called nuclear fusion, and it's an attempt to replicate the power of the sun, no less. It is worthy of mention because in pursuit of this goal the major governments of the world have joined together to spend €10bn on the Iter project.

The rationale for attempting nuclear fusion rests on, to the layman, a bizarre fact of physics. Namely, while the splitting of an atom in nuclear fission can release the enormous energy that binds the atom's particles,

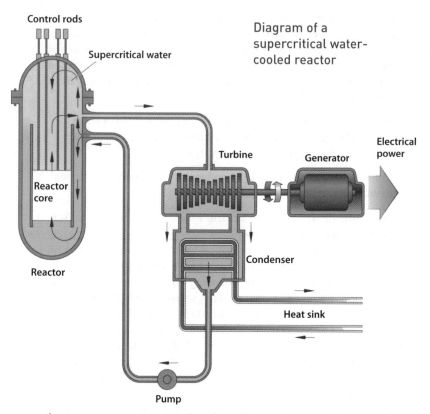

Diagram of a supercritical water-cooled reactor

there is even more energy to be released by combining or fusing certain atomic nuclei (of hydrogen) into a bigger nucleus.

What's more, nuclear fusion carries neither of the two main risks posed by fission. First, while the radioactivity of split atoms is long-lived after nuclear fission, in the fusion process it is limited to tritium, a radioactive form of hydrogen which decays quite quickly to become innocuous.

Second, the fusion process is far less likely to spin out of control. The fission process leads to particles of a split atom splitting other atoms in a chain reaction that has to be carefully controlled in a power reactor – otherwise you got an uncontrolled reaction, otherwise known as a nuclear bomb. So far as we know, there is relatively little danger of this happening with fusion. The fusion process uses tiny amounts of nuclear fuel. Moreover, there is little chance of a runaway reaction, mainly because the fusion reaction is so fantastically difficult to achieve and maintain in the first place.

Therein, however, lies the problem with fusion in terms of generating electricity. Making a fusion bomb – popularly known as a hydrogen or H-bomb – is relatively easy. It is done with the very blunt instrument of using a fission explosion to set off a fusion explosion that only has to be momentary. This technique is hardly appropriate if one day you want to be able to use fusion to generate electricity for customers with any continuity. So we have the odd situation that less than ten years after the first fission bomb we had fission-produced electricity, but that more than fifty years after the first H-bomb, we are still decades away from obtaining useful electricity from fusion.

Fusion is an attempt to replicate the way that the sun and stars fuse hydrogen nuclei. But we poor earthlings can't do it in quite the same way. The sun and stars have huge gravitational pressures upon their nuclei, which mean they fuse at the relatively low temperature of a mere ten million degrees centigrade. In the absence of such pressures on earth, fusion requires temperatures as high as a hundred million degrees centigrade to overcome the repelling forces of positively-charged hydrogen nuclei.

How do you go about achieving such high temperatures? One of the chief difficulties is how to contain the superheated fuel particles without letting them touch any solid container that would just leach away the heat. The main solution being pursued uses a tokamak, a Russian-designed doughnut-shaped machine that produces a magnetic field to contain the hot fuel plasma, and thereby keep the fuel particles away from the tokamak's walls. The containment problem is not just one of preventing the fuel particles hitting the reactor walls and dissipating heat – it is also to keep the particles together long enough for the fusion reaction to produce energy at a rate greater than that of the energy input.

The Iter consortium

Pursuing this solution is an international consortium known as Iter, made up of China, the EU countries, India, Japan, Russia, South Korea and the US. Originally Iter's name was an acronym standing for International Thermonuclear Experimental Reactor, but some people were understandably nervous about the word thermonuclear being used in close conjunction with the word experimental. So sometimes the word tokamak is used instead of its place. Iter means "journey" in Latin – and fusion is certainly a journey to an unknown destination.

Iter is building on previous experience with tokamaks – notably the work achieved by an EU project, the Joint European Torus (JET), at Culham near

Oxford in the UK. Construction of the Iter reactor began at Cadarache in southern France and is likely to take almost ten years to complete. The goal of the project, which in 2009 had a projected cost of €10bn, is to produce power of up to 500 megawatts for up to 1,000 seconds – as compared to JET, which only produced 16MW for less than a second.

In view of the incredible demands of high temperature fusion, it was not surprising that when two scientists, Martin Fleischmann and Stanley Pons, announced in 1989 that they had achieved nuclear fusion at room temperature, they created a sensation. Fleischmann and Pons reported producing nuclear fusion in a tabletop experiment involving electrolysis of heavy water with a palladium electrode. They suggested that this had produced excess heat on a scale that only a nuclear procress could explain. They also reported a small amount of the by-products of a nuclear reaction. Unfortunately cold fusion turned to confusion, when other scientists tried, but failed, to replicate the Fleischmann-Pons experiment. It was ultimately concluded that the excess heat the two scientists reported was the result of a chemical reaction, not nuclear fusion.

Despite being a member of Iter, the US government is hedging its bets. It is funding research into an alternative form of hot nuclear fusion using lasers and has set up an institution called the National Ignition Facility at the Lawrence Livermore laboratory in California. The central experiment

The Joint European Torus tokamak – with overlay showing it in operation.

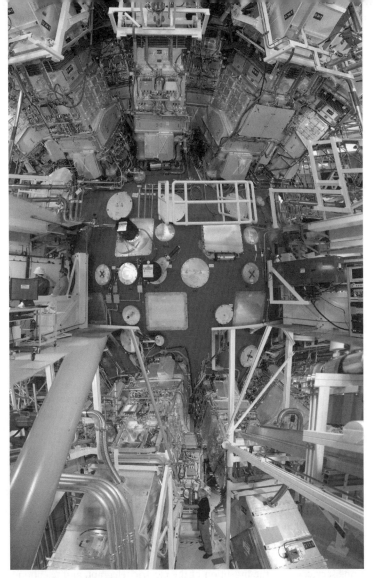

A composite photograph showing all three floors of the target chamber in the US's National Ignition Facility.

focuses a series of very powerful lasers on a pellet of frozen hydrogen in order to heat and pressurize it to the point of fusion. But for all the effort and expense that governments are devoting to it, nuclear fusion remains so uncertain a prospect that nobody puts it into any projections of future energy sources.

Extending the electrons
New things in electricity

Electricity is not a fuel or a primary energy source (except in the uncontrollable form of lightning, that is). But its role in the energy infrastructure is crucial, as it is the most convenient means of transmitting energy.

The basics

Primary fuels (oil, gas, coal, nuclear and wood, for example) can heat up water, turning it into steam, which can then rotate the blades of a turbine, which in turn can spin a generator. In that generator the relative rotary motion of magnets and windings of wire (usually made of copper) produce electricity. This electricity can then be used to power motorized appliances simply by reversing the process – turning electricity back into mechanical energy. In the case of solar photovoltaics, sunlight is converted directly into electricity.

A constant theme in energy history has been the steady advance of electricity into new uses. In the nineteenth century, electricity began to displace gas and candles in lighting, and created entirely new communications services with the invention of the electric telegraph and telephone. In the twentieth century, it has replaced human muscle power in washing clothes and dishes, and provided new services in refrigerating food. And in the twenty-first century, it is powering electronic devices such as computers, which hold open the possibility of rationalizing supply and demand better than ever via smart grids.

Grids, our national electricity networks, are themselves, of course, the result of this steady progress in electrification. In the nineteenth century, electric lighting systems or motors required their own batteries or generators because there were no networks. Then local electricity networks were developed which appliances could tap into, and eventually these were integrated into national grids.

Electrification came to be seen as vital to a country's progress. In 1920, Lenin famously defined Communism as "Soviet power plus electrification". In 1936, Franklin D. Roosevelt created the Rural Electrification Administration as an integral part of his New Deal for America. And today, one of the most universally recognized measures of global deprivation is the fact that two billion people still lack access to electricity. On the basis of present trends and policies, the IEA estimates that 1.3bn people will still be without electricity by 2030, but that universal access to electricity could be achieved by that date with the relatively small extra investment of $35bn a year and minor increases in emissions. (In the table below, note China's outstanding success in providing access to electricity for more than 1.3bn of its citizens. Of course, that very success compounds its emissions problem.)

The power-less in 2008

Country/region	Population without power (millions)	Electrification rate in urban areas (percent)	Electrification rate in rural areas (percent)
North Africa	2	100	98
Sub-Saharan Africa	587	57	12
China	8	100	99
India	405	93	53
Other Asia	396	85	48
Latin America	34	99	77
Middle East	21	99	70
Total	1,453	90	58

Source: International Energy Agency, *World Energy Outlook 2009*

Electrification: its impact on energy (and climate)

The explosion of electrification over the twentieth century was a welcome trend; electricity can provide an extraordinary array of services. It is now far more than getting light, warmth, refrigeration or motion at the flick of a switch. The Internet has now connected up the world, revolutionizing communications. Further electrification of our energy systems ought to be something to welcome.

However, in terms of climate and resources, our increasing electricity usage has a cost. Could further electrification of our lives be environmentally beneficial? Yes, if the electricity is produced by low-carbon means – which, in the world we currently live in, essentially means using renewable or nuclear energy sources. It is important to ask what impact electricity, itself the physical product of the energy system, is having on that same energy system.

Adaptation to climate change

One example of this is air conditioning. An inevitable result of global warming will be the increased use, for those who can afford it, of air conditioning. For some southern states of the US and of the European Union, electricity consumption now peaks in the summer, not winter. There is a vicious circle developing here. Energy use creates some 65–80 percent (depending on how industrialized a country is) of the greenhouse gases creating global warming. Responding to a symptom of climate change by upping our use of its causes will clearly worsen the problem.

Yet electrification will play a vital part in mitigating climate change. In terms of transport, it has the potential to decarbonize a considerable amount of car travel (see p.230) via battery-operated electric cars and plug-in hybrid electric vehicles. It is electricity or rather micro-chips and electronics that can provide new ways of interacting with people, such as tele-working or tele-conferencing, cutting down on the carbon footprint of international travel.

And, as we will investigate in this chapter, increasingly "intelligent" and interactive approaches to our electricity supply could make a significant global difference to emissions, through smart grids (see p.104) designed to create a better real-time match between supply and demand – shifting peak demand to off-peak times and so making it easier to satisfy.

Saving it on the old, blowing it on the new.

Enormous progress has been made in recent years in increasing the energy efficiency of traditional household appliances. This is true even in the US, where tax and price pressures have hardly been the strongest. For instance, the average American refrigerator in 1972 was around 18 cubic feet in size and consumed 1986 kilowatt hours a year; by 2001 the average US fridge, whose size had by this time grown to 24 cubic feet, was consuming 1239 kilowatt hours a year. Despite increasing by a third in size, and upping its efficiency, the price of the average fridge had come down thirty percent over this period due to technological improvements.

However, much of the electricity saved on traditional items has been blown on consumption of the new goods and services made possible by the digital revolution.

Televisions

According to the International Energy Agency, more TV sets are being sold around the world than ever before, at a growth rate of 3.7 percent a year. This is faster than growth in the number of households gaining access to electricity (two percent a year), so evidently the number of TV sets per household is rising globally. Furthermore, liquid crystal display (LCD) panels and plasma screens have fast replaced the much older cathode ray tube (CRT) screens, and generally speaking, these screens are much larger. The Energy Saving Trust has pointed out that plasma-screen TVs typically consume four times as much energy as their predecessors.

Computers

Nearly one billion people use computers, and computer sales are still rising. Not only is the number of households with a computer growing, but the number of personal computers is also increasing. Increasing broadband Internet access has, understandably become a priority government in many countries. But broadband has also encouraged people to spend much more time on the Internet – more than watching television and much more than reading newspapers or magazines.

Standby mode

The amount of electricity consumed as standby power for products is growing. Of all the avoidable expenditures of energy resources, this is perhaps the most absurd, serving no useful purpose. That hasn't stopped standby power – defined by the IEA as "power required by a product when not performing its primary function" – becoming a standard feature on many products.

Remote-controls or time-clocks on products require power in order to respond to start-up signals and to keep clocks running. TV set-top boxes also often need to stay in high-power mode so that TV service providers can send security and software updates. But other products clearly ought to have devices enabling them to automatically power down.

Smart electricity

One technological concept that could revolutionize electricity is the smart grid. Today's power grids supply electricity in a unidirectional flow – from centralized power stations to millions of customers. Tomorrow's grids will use digital technology to make our electricity supply more reliable – capable of integrating numerous and intermittent sources of renewable power, be they solar, wind, tidal or hydroelectricity. But they will also enable customers – in two-way communication with suppliers or grid operators – to re-shape their demand for electricity so that it saves them money and fits better with supply.

In other words, the smart-grid idea turns a rather passive system, driven by suppliers alone, into a more dynamic, intelligent system capable of reflecting the interests of customers as well as suppliers to their mutual benefit. This is the goal now being pursued in many developed countries. The European Union has had a smart-grid technology programme going since 2005 (though the pace of roll-out for the technology varies considerably among EU states). In the US, Barack Obama's administration made smart-grid technologies a priority of the $80bn devoted to clean energy in its 2009 Recovery Act.

Micro-generation

Today's grids are already having to adapt to renewable power. One factor is their ability to deal with potentially abrupt surges in power – when the wind blows or the sun shines for example – and, when such surges subside, for drawing on compensating sources of power. Integration of different types of power will be key. This task will become more complex as the percentage of renewable electricity in the national system increases, and as the number of providers of that renewable electricity increases, with the development of decentralized energy generation. This decentralization refers to the new opportunity provided by smart grids for businesses, communities and even individuals to provide heat and light to satisfy their own energy needs – and, if they generate any surplus power, to sell back the excess to the grid. This energy generation will typically be from solar panels on roofs, wind turbines on the roof or on the ground, wood boilers or from a variety of heat pumps.

Such electricity generation is often referred to as micro-generation. In the UK there are now over 100,000 micro-generation units, though the vast majority just produce heat and/or light merely for themselves and without transmission through the grid. In Europe, sales of power to the grid tend to get a fixed feed-in tariff as an extra incentive. The UK is planning to introduce such feed-in tariffs from 2010.

Micro-generation adds to the diversity of supply through wind, solar and biomass power to the grid. It therefore, in theory, increases a country's home-grown (or home-blown) energy security. Short-term energy stability, however, is more of a problem for the grid if, for example, a sizeable number of wind farms were stopping and starting in unison. So in order to check stability, another smart-grid technology has been developed using devices called phasor measurement units. These are high-speed sensors, distributed in large numbers across a grid, which sample the current and voltage several times a second, and display waveforms of them. These waveforms show up any discrepancies, enabling electricity companies and grid operators to react to any interruptions or blackouts on the grid at great speed.

A different, constant problem of sending electricity long distances along a wire is the transmission loss of energy , which can be as much as ten percent of what was originally generated. Some of the electricity is always wasted heating up the wires, which to varying extents resist the flow of the electrical current. The bigger the electrical current, the more of it will be lost.

One solution has been found: to reduce the *level* of electrical current, but to increase the *force* behind it – the voltage – to get the electrical current to its destination. This has been applied in the form of high voltage direct current (HVDC). Most long-distance electricity transmission is currently achieved using high voltage alternating current (HVAC). The majority of electricity comes in the form of alternating current AC which switches direction back and forth many times a second and which drives almost all electric motors and appliances – in contrast to direct current DC electricity which flows continuously in one direction. Alternating current has been the preferred option for most dispatch of electricity around the grid, because it is easier to adapt to different voltage levels. But HVDC is better for transmitting large amounts of power over long distances with lower capital costs and lower transmission losses than HVAC.

Therefore HVDC power transmission may become more important in a low-carbon economy that is increasingly electrified via renewable power coming from distant parts (see Desertec project in Solar power section in Chapter 6). Transmission loss can be nearly eliminated by using superconductor cables, but this is costly. For example the cable is far more complex to manufacture than copper wire, and the cable has to be cooled. Since it will probably be impossible to ever reduce transmission losses to zero, there is one clear advantage in micro-generation that is unlikely to

Stacks of thyristors – semiconductor devices capable of switching power on a megawatt scale – hang from the ceiling of a "valve hall". Thyristors lie at the heart of high voltage direct current (HVDC) conversion, and the ones pictured here are part of a joint project between Siemens and BHEL to supply power, via HVDC cables, to the area around New Delhi in India.

ever be rendered redundant, having the generation of electricity at or near the point of its consumption.

Demand-side

But the really exciting aspect of smart grids is on the demand side, with the possibility of changing consumer behaviour in order to tackle the problem of peak demand. This problem is uniquely difficult in electricity, because you can't store it in any significant way. Supply and demand must match – electricity has to be consumed the moment it's generated, and the moment it's demanded it must be produced.

In order not to be caught short by some unexpected peak in electricity demand, utilities need to keep some back-up generators spinning away, ready to be hooked up to the grid at any moment, but which might be used only rarely. In the UK, according to a recent Brattle Group study, nearly seven percent of generation capacity is used a mere one percent of the time during a year, at moments of highest peak demand. Building and maintaining generators that produce electricity for less than a hundred hours a year is a waste of money, and operating these generators as spinning reserve is a cause of needless greenhouse gas emissions. So, to remain with the UK case, if peak demand could be lowered during this one per-cent of the year or shifted in time, then the UK could avoid building that extra seven percent of generation capacity. This would make a much more efficient system: money and emissions would be saved.

But how do you go about anticipating, lowering, shifting or re-shaping the power demands of an entire nation's households and offices? That is where smart meters come in. These do what ordinary gas and electricity meters do today – measure consumption in order to bill the customer. However, smart meters do not require electricity or gas company personnel to turn up in person at a house or business to read the meter. Smart meters can transmit up-to-date consumption figures to be read electronically either to utility trucks going up and down streets or directly back to the supplying energy company.

This is why, even in the UK, which is a latecomer to smart metering, energy companies are beginning to advertise an end to estimated bills, which often arrive through the letter box months after you've used the electricity and are frequently inaccurate. So far the only country to roll out this sort of smart meter on a national scale is Italy. While it was still Italy's state electricity company, Enel decided on a national campaign which installed 27 million smart meters between 2002 and 2005. The company said that, with annual savings of €500m – there were no longer any house visits by meter readers, and less customer fraud – it could recoup the €2bn cost of the meters in four years.

Elsewhere in Europe and in North America, deployment of standard smart meters has been patchy. The UK only plans to have national coverage of smart meters completed by 2020. But the UK is also conducting trials – in some 18,000 households during 2009–10 – into how consumers respond to real-time energy display devices to change their behaviour.

Behaviour change is the key. For smart meters can give the customer hyper-detailed, up-to-the minute price and volume information with which to make a choice about which appliances to use. If smart meters are to influence consumers to change their electricity consumption pattern, there has to be a parallel smart *pricing* of electricity. Intelligent pricing of this kind could, for instance, provide an incentive to people to run their dishwasher at night, when demand and prices are lower.

This smart pricing is usually referred to as time-of-use (TOU) pricing.

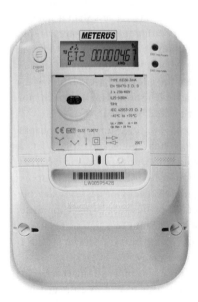

This European smart meter uses two-way communication to connect remotely and interface with gas and water meters.

But it would be really boring to go to the meter every half an hour to check the price of the electricity you're using. So some smart meters can control appliances remotely, in accordance with guidelines pre-set by consumers. Smart meters can automatically shut off washing machines and dishwashers when the TOU price goes above a certain level. Smart thermostats can automatically raise the temperature setting on an air-conditioning unit by a couple of degrees when the price rises above a set threshold.

So far it is mainly in the US where smart meters of this sophistication have been tried out. One example is the city of Boulder in Colorado, where the local utility company Xcel, in alliance with several IT companies, is equipping ten thousand homes with devices to control electric appliances remotely. Not all trials of smart meters have gone down well with consumers, who sometimes see smart metering as just enabling energy companies to cut personnel costs and to make it easier for the companies to cut off power to customers in arrears on bill payments.

The responsiveness of people to energy-pricing incentives is also often overestimated – look at the way people go on buying petrol no matter what the price is. But overall the trials show, according to US Department of Energy officials, that smart metering can cut peak electricity demand by as much as fifteen percent, by either shifting it in time or reducing it in volume.

Renewables
On the rise, but an uphill struggle

Renewable energies are the most essential part of the low-carbon economy. Although investment in renewables took a big hit in the 2008–09 economic recession, their share of the world's energy mix is rising. Increasing their use, however, remains an uphill struggle, because in almost every case renewables require government support or subsidy to compete with fossil fuels. But what are renewable energies? And why are they harder to generate, both technically and commercially, than fossil fuels?

The basics

By definition, renewable forms of energy renew or replenish themselves naturally. Unlike oil, gas, coal or uranium ore for nuclear power – the reserves of which have not increased since their creation in distant geological time – we can keep using and re-using solar, wind and tidal power, and growing crops for biomass (with the proviso of access to adequate water). There is no finite stock of these renewable energy sources. Moreover, using them over and over again is sustainable.

These renewable energy sources are clean. They produce none of the by-products of fossil fuels or nuclear power, such as carbon emissions or nuclear waste, which damage the climate and make life harder for future generations. Renewables and the low-carbon economy are key to achieving the wider goal of energy sustainability. But for an energy system to be sustainable, supply and demand have to be in balance – and this requires not only a cleaner supply of energy, but also greater efficiency on the demand side in using energy.

Before examining each renewable in detail, it is useful to point out one broad distinction that policy makers increasingly make.

Old renewables

These are already big contributors to the global energy balance, but for various reasons there is no widespread desire to increase them. One of these old renewables is traditional biomass – the firewood, animal dung and farm waste that has been gathered by rural households since time immemorial for cooking and heating. This is thought to amount to some seven percent of the total global energy mix. But smoke from this kind of biomass can be a killer in unventilated kitchens. Firewood collection is also fundamentally unsustainable when it leads to deforestation.

The other main old renewable is hydroelectric power, which is by far the most significant renewable source of world electricity generation, accounting for much as sixteen percent of it. But, with the exception of Asia (see hydropower section), big hydroelectric dams have fallen out of favour, because of the displacement of populations and disruption of fish stocks they generally cause. Still in favour are smaller hydro projects, especially run-of-the-river projects, which are capable of funnelling some water into turbines without blocking or damming whole rivers.

New renewables

Energy sources that fall into this category include solar, wind, tidal and wave power, and the environmentally controlled cultivation and use of biomass for power and heat, and of biofuels for transport. The contribution of these new renewables is tiny – around 0.2 percent of global primary energy supply. In fact, in the case of windmills and a few tidal power schemes, these renewables are not new and indeed go back centuries. But what is new about them all is the new enthusiasm to develop them.

Renewable energy 2006
(inc. hydropower; exc. traditional biomass)

Contribution to total energy demand	7%
Contribution to total electricity demand	18%
Contribution to total heat demand	6%
Contribution to total transport demand	1%

Source: International Energy Agency, *World Energy Outlook 2008*

So the growth in renewables is mainly coming from the new renewable energy sources from the sun, the ocean and, above all, wind. But this growth has had to be government-supported in almost every instance.

▶ **The 27 member countries of the European Union** have set themselves mandatory national renewable energy targets for 2020, averaging out as a 20 percent renewable share of total EU energy.

▶ **Thirty US states** have differing "renewable portfolio standards" that require a minimum (ranging from ten to thirty percent) of their electricity to be renewably generated.

▶ **China has a 2006 Renewable Energy Law,** providing for the compulsory connection of all renewable power plants to the grid and the compulsory acceptance by utilities of all renewable power offered for sale.

But why, you may well ask, should renewables need such support to compete with and hopefully displace fossil fuels? The answer is that there are some generic obstacles that renewables must overcome with regard to fossil fuels – and before examining individual renewables it is worth pointing some of them out.

Today, only 29 percent of Nepal's forest-cover remains. The main reasons for deforestation there have been land clearings for agriculture, the demand for timber and, especially, the need for firewood. About 87% of domestic energy in Nepal is produced by firewood, used for cooking and, during winter, for heating.

Stumbling blocks for renewables

Financial costs

Renewable energies have a very high ratio of capital equipment costs to the cost of fuel. In other words, the cost of setting up a renewable power plant – the wind turbines or the solar arrays, for example – is vast when compared to what will be powering it. In the case of solar or wind power the fuel cost is zero, and in the case of waste for incineration, the fuel cost may even be negative (if renewable generators are paid to burn waste).

Over the long term, this will be a relative advantage for renewable energy producers. Their capital costs are likely to come down over time, as economies of scale in technology and manufacturing kick in, whereas the cost of fossil fuels, which will become harder and harder to explore and exploit, is likely to rise over time. In the short term, however, it is a relative handicap for renewable-energy generators that virtually the entire cost of their operations has to be paid up-front in buying the equipment. This was one reason why renewable investment was hit disproportionately hard in the 2008–09 recession.

Energy content and convenience

Energy density – or how much energy a fuel carries – is measured by the ratio of the energy stored relative to the mass of the fuel it is stored in. By this measure, the energy stored in an atom far outstrips anything else. But fossil fuels – coal, oil and even natural gas – come at least second.

Most renewables, in the energy stored in proportion to the mass of wind or tidal current storing it, come in a distant third. Convenience of transport is related to density: coal and oil are the densest fossil fuels and the most transportable; gas is the most dispersed and the least convenient fossil fuel to transport. Most renewables can only be turned into electricity, which can then be transported – though biomass can, at considerable cost, also be converted into liquid or gas.

Infrastructure

A problem common to all renewables is the need to rethink infrastructure optimized over the past century for fossil fuels. Because renewables are a less dense form of energy than fossil fuels, they generally require a more extensive production system.

They need more space to grow (in the case of biomass) or to be collected (wind and solar) than coal, oil or gas, which contain energy in a more compact, and – crucially, for environmental reasons – underground form. So wind farms with turbines a hundred metres high, or solar arrays with a couple of square kilometres of reflectors, will usually take up more space, and are often more visible, than a coal mine, oil and gas field. Not that coal slag-heaps, gas flares or oil derricks are exactly invisible, of course.

Moreover, it is not just that some renewables operations are more extensive and obtrusive than the hydrocarbons they might replace. They are also usually in different places to conventional energy sources, and may have different logistical requirements. The best wind power is normally to be found in remote places, requiring the building of new links to the electricity grid. A good example of this is Scottish wind farms: northern Scotland produces the best wind in the UK, but southern England is where most of the UK's consumers are.

New types of infrastructure may have to be used. For instance, a power plant that uses coal can bring it in on a single-rail track or by barge on a canal. A plant using biomass, however, will have to have it trucked in from a far wider catchment area. Rob Church of the American Council for Renewable Energy (Acore) makes the point that the economic dice are loaded in favour of hydrocarbons. "If we were starting from zero, it would be OK. But we have laid an infrastructure for coal already … many of the problems with renewable energies are because we are having to change delivery systems as well."

Intermittency

The other characteristic of wind and solar power is unpredictable intermittency (in contrast to tidal power, which is predictably intermittent). As a grid operator, you have to take the power when you can get it, and make other arrangements when it is not.

Wind and solar power is, in the jargon of the electricity industry, "non-dispatchable". A grid operator cannot just order up some wind or solar power, as they might coal or gas, and dispatch it to the grid. Rather, the operator has to take the wind power when the wind blows and solar power when the sun shines. And make other arrangements when they're not available.

Renewable energy gets going – in FITs and starts

There are two common ways of subsidizing renewable electricity. One is to give renewable electricity producers guaranteed prices in feed-in tariffs (FITs), and the other to impose on distribution companies an obligation to buy a certain quota of renewable power, in conjunction with tradeable green certificates. Both schemes act as a subsidy, the cost of which ends up on the energy bill of the consumer.

FITs have proved more effective in raising renewable electricity production (for example in Germany, Spain, Denmark) than quota obligations (which the UK has mainly relied on). So about two-thirds of the European Union's 27 states now rely on FITs. The UK still operates a Renewable Obligation Certificate (ROC) scheme requiring energy companies to get a rising share (9.7% in 2010) of their total supply from renewable sources. But the UK introduced a FIT in 2010 for small producers of renewable energy. In the US no federal subsidy scheme exists for renewables (except for bio-fuels), but some thirty states have a quota obligation, usually referred to as Renewable Portfolio Standards, and a few US states are moving to FITs.

FITs are increasingly popular because they give producers greater financial certainty, and can be set at different levels for different technologies. Germany, for instance, has become a world leader in solar technology by setting a high FIT for solar PV – a feat for which it has been praised in technical circles, but criticized in financial terms. A FIT gives renewable producers a fixed guaranteed price for however much power they feed in to the grid; some countries operate a modified FIT system that gives producers a fixed premium payment to top up the electricity market price. Quota obligations provide less financial certainty. Renewable energy producers sell their green power for whatever they can get in the electricity market, and they also sell accompanying green certificates (ROCs in the UK case) to suppliers who need the certificates to show they have fulfilled their quota obligation. In this system, renewable producers' income depends on two fluctuating values, the market price of green certificates and the market price of electricity. A further problem would come if ever the quota were exceeded, causing over-supply of green certificates and a drop in their value.

One solution is demand-management, or load-shedding. This involves companies and/or individuals agreeing to take power cuts in return for a payment or discount and is discussed in tandem with smart grids in the previous chapter (see p.106). But there are plenty of other options.

"Electric Mountain", the Dinorwig power station in Snowdonia, is the only non-fossil-fuelled power station in the UK that could give the UK a "black start" – re-starting the national grid in the event of its failure.

Importing more power into the system

The traditional fall-back has been gas-fired power, which is relatively cheap to build and therefore tolerable to keep in reserve for moments of peak demand. But the green ideal would be to have renewable back-up.

If the system is big enough, then it is sometimes possible to balance a drop in wind power in one part of the system with a rise in wind generation in another part of the system – for instance a lull in southern England compensated by rising wind power in Scotland. But even in a big state such as California this has not worked well – the wind or sun either blows or shines throughout the state or not at all. Furthermore, transmission over long distances leads to electricity losses. Sometimes wind and solar power can complement each other well. It is, for instance, an awkward fact for wind that an anti-cyclone high pressure, with little or no wind, can bring the coldest and the hottest weather. Fortunately, such anti-cyclones are usually sunny as well as windless, and produce solar power.

Storing electricity

So far there is only one type of storage – that of hydroelectricity – that works well as a back-up to intermittent wind or solar. Denmark depends on wind power for as much as twenty percent, on average, of its total energy consumption, and it can do so because in the event of a shortfall, it can draw on Norwegian and Swedish hydropower. In many other countries, pumped storage is very common. This requires two bodies of water: a higher and a lower lake (the lower one could, in fact, be the sea). The process involves pumping water from the lower to the upper lake at times when electricity is cheap (usually at night) and then releasing it from the top lake top in times of need to turn turbines and generate power.

The UK has one unusual pumped storage facility, the Dinorwig power station in Snowdonia in Wales. It is unusual in that the top part is a hollowed–out mountain, a process that took ten years (1974–84) to complete. It was initially designed, before the days of commercial wind power, purely to provide a store for cheap night-time power from the UK nuclear power plants. But now there is an extra rationale for Dinorwig as a back-up for wind power. Dinorwig also has the capability of re-starting the UK national grid after a complete power failure, were that ever to happen.

Wind power

The answer, my friend...

Of the "new renewables", wind has grown fastest. Since 2000 the installation of wind power capacity has quadrupled, doubling about every three years. Wind – at least onshore wind – has also become the most mainstream of renewables. The technology is well understood (see below). Its economics are improving, its costs are coming down, and it forms part of the portfolio of virtually every big energy company.

Europe pioneered the modern wind power industry. Manufacture of modern wind turbines started in Denmark, which now generates twenty

A wind turbine manufactured by Suzlon, one of the developing world's big success stories in renewables manufacturing: Suzlon is Asia's largest wind turbine manufacturer, and is the fifth-largest in the world.

percent of all its energy with wind, the highest proportion of any country. Denmark also pioneered the offshore wind industry.

In volume terms, bigger European countries – Germany and Spain – have moved ahead of Denmark in recent years, only to be overtaken themselves by the US: in 2008 the US overtook Germany in nameplate capacity of installed wind power. Together the US and Europe account for eighty percent of global wind capacity – both regions have good wind resources and the money to exploit these resources. But China and India are, respectively, fourth and fifth in the world wind-power league, and India is home to Suzlon, one of the big international makers of wind turbines.

Scale counts

Generally, wind speeds rise as altitude increases and obstructions diminish. So the taller the turbine tower, the greater the wind speed it will capture. And there is a fact of physics which shows that this matters – the power in the wind is proportional to the *cube* of the speed of the wind. So a turbine which will produce 625 watts at 15 mph will not just produce double the power when the wind speed is doubled to 30 mph, but eight times the power, at 5000 watts or 5 kilowatts (KWs).

The size of the turbine rotor also counts disproportionately. This is because the power is also proportional to the *square* of the turbine diameter. So a doubling of the turbine diameter quadruples the increase in power. The amount of electricity from wind surges dramatically as the wind gets faster and as turbine blades grow longer: it has the effect of even further increasing the variability of wind-generated electricity.

How a wind turbine works

▶ **Step 1** The wind turns the blades

▶ **Step 2** The blades turn a shaft inside the nacelle (a box at the top of the turbine)

▶ **Step 3** The spinning shaft is linked to a generator

▶ **Step 4** The generator uses magnetic fields to turn rotation into electrical energy

▶ **Step 5** The electricity goes to a transformer that adapts its voltage to the grid

Nysted Offshore is currently the world's largest offshore wind park, with 72 windmills and a total capacity of 165.6 megawatts.

This explains why, however laudable it might seem to put a micro-turbine on the roof of your house, it actually makes more sense for individuals to pool their efforts and invest in wind-power cooperatives, so as to get some economies of scale with taller turbines. For micro-turbines are unable to capture any decent level of wind unless they are more than ten metres clear of a house roof, which in most cases is impossible or restricted by local planners.

All at sea

One of the reasons for erecting wind turbines offshore is to achieve a higher wind speed (the other important factor being to avoid the environmental objections that onshore turbines attract). It is also easier to transport big components at sea. However, fixing the turbines to the sea

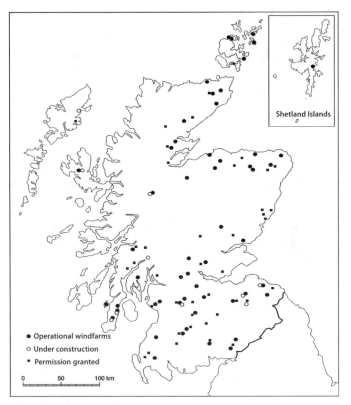

Shetland Islands

• Operational windfarms
○ Under construction
▪ Permission granted

0   50   100 km

Not all of the UK is NIMBYish about onshore wind power: Scotland has seen a boom in wind-generated electricity, as the above map shows.

floor can be a problem, though obviously this is easier in shallow water. Europe has taken the lead in offshore wind partly because it has a lot of fairly shallow and windy water (up to 25m metres) around the margins of the North Sea. For its part, the US has some shallow water with good wind resources off its Atlantic coast. The Gulf of Mexico also has a lot of shallow water, but it's unworkable as a location because the wind there tends to come in hurricanes or not all.

Solar energy
Here comes the sun

Most forms of energy come from the sun. The notable exceptions are geo-thermal energy (drawing heat from the earth's core) and tidal power influenced by the moon. All plant life depends on the sun; fossil fuels are plants that have fossilized into fuel; and biomass is biological material. These are indirect sources of solar energy. But of course there are more direct sources of solar energy – three in particular.

Solar thermal

The sun's rays have been used for centuries to heat water – in this sense solar thermal energy is nothing new. The most common modern process of solar thermal technology involves heating tubes containing a liquid which is pumped into a heat exchanger, which transfers the heat to a hot water tank. It is a fairly simple technology, the use of which has increased, but not to the extent that it deserves. It has, however, been taken up enthusiastically by countries with sunny climates which also lack energy alternatives – such as Israel, Cyprus and to some extent China. In the US, the majority of solar thermal energy is of the low-temperature variety – panels that are used mainly for heating swimming pools.

Solar panels used for heating water in Santorini, Greece.

A Mauritanian villager inspects a solar panel near her local well in the Sahara desert. The panel uses solar energy to generate electricity to pump up water from the well.

Solar photo-voltaic (PV)

PV technology turns sunlight directly into electricity. The light energy of the sun (or "photon") displaces electrons embedded in solar cells – slices of silicon – and sends the electrons across an electrical junction, thus creating a voltage.

The prospect of being able to turn the universe's predominant natural source of energy – the sun – directly into electricity without the fuss of any intermediate steps is so enticing that the growth rate in solar PV production has been truly extraordinary. According to the International Energy Agency's *Renewable Energy Statistics* report of 2008, investment in PV within OECD countries rose from 19 GWh in 1990 to 2626 GWh in 2006, an average yearly increase of 36 percent. Much of this was driven by a very high subsidy or feed-in tariff in just one country, Germany, where PV capacity increased from just 1 GWh in 1990 to 2200 GWh in 2006, an average annual growth rate of 62 percent.

Concentrated solar power (CSP)

This technology uses various techniques to focus the sun's rays onto a liquid that is heated up enough to boil water and produce turbine-spinning steam in the usual way. It typically utilizes reflective dishes or troughs that focus the sun's rays on a single point to get maximum heat.

The process is simpler than solar PV, but the process of collecting the sunshine has to be done in quite extensive sun farms, comparable to wind farms. Desert areas are therefore the most appropriate. In the Mojave desert in the southwest United States, CSP has been produced since the 1980s, with arrays of parabolic mirrors used to heat up troughs of liquid. More recently, in Europe, a different CSP design has been tried out – that of solar "power towers". In 2007, the first solar tower started operation near Seville. It consists of several hundred mirrors placed in a circle around a tower so that they all reflect sunlight at a single point at the top of the tower. This concentrated sunlight turns water into steam and so drives an electricity-generating turbine. However, this spectacular power

An array of solar mirrors directly concentrate the sun's energy on the boiler atop a power tower in the Mojave Desert.

tower, which appears bathed in a heavenly light, only generates around 11 megawatts, enough power for about 6,000 homes.

A far more ambitious scheme, called Desertec, has been conceived by a coalition of mainly German companies to tap the solar power potential of the Sahara and transport it across the Mediterranean to Europe. The consortium behind the Desertec Industrial Initiative, which was launched in 2009, believes that CSP arrays in North Africa and the Middle East could provide up to seventeen percent of Europe's energy needs by 2050. Crucial to the success of the scheme would be carriage of the electricity via high-voltage direct current (HVDC) cables, which lose less current in transmission and are better for underwater transmission than high voltage alternating current (HVAC) cables.

Desertec is backed by German engineering, energy, insurance and financial companies, which are keen to see their country's home-grown solar expertise in more sun-blessed regions. The scheme has also been taken up by European politicians on the look-out for promising partnership projects with North Africa, which are generally rare. But although the Desertec consortium does contain one Algerian company, there does not seem to be much enthusiasm for it from North African governments, which may see Desertec as undermining their oil and gas exports to Europe.

What makes solar energy interesting is its versatility. It can be used directly for heat without going through the intermediate step of being converted into electricity, as wind power must be. But through the PV process, solar energy can also be transformed directly into electricity without the need for any big generating mechanisms (such as wind turbines) or any transmission infrastructure. So solar PV is, theoretically, ideal for use in the middle of cities and in the middle of the bush far from any grid.

Moreover there are no laws of physics that make micro-solar PV inherently less efficient than macro-solar PV, in the way that big wind turbines with big rotor arms capturing faster winds are disproportionately more efficient than micro wind turbines. So, solar PV has very good small-scale potential, while CPS has large scale potential when it comes to making use of desert areas that serve no other purpose.

Water

Making waves

Water does not travel at the speed of wind. But for the same volume, it has far greater force because it is more than eight hundred times denser than wind (which is also why tidal and wave machines have to be much stronger than wind turbines).

Ocean power

The power of the sea is immense, and endlessly renewable. It comes in essentially two forms. There is the power of the tides, resulting from the gravitational pull of the moon and to a lesser extent the sun, and of the earth's own rotational spin. And there is the power in waves, themselves created by wind that is a by-product of solar power.

Yet speed has the same disproportionately dramatic effect on water power as it has on wind. Water energy rises exponentially with water speed – so that a tidal stream moving at sixteen kilometres an hour will have *eight* times the force of a current moving at half the speed, at eight kilometres an hour. Moreover, tidal power may be spasmodic, but it is also predictably spasmodic (which is unusual for a renewable energy source). Consult your tidal charts, and you will know months or years ahead when surges in tidal power will occur.

In contrast to tides, but in common with wind, waves are unpredictably intermittent. But waves are not intermittent on precisely the same cycle as wind, which should be an advantage in terms of evening out the flow of electricity from the two sources. Because they are caused by wind, waves take a certain time to swell up after the wind begins to blow, but also continue to roll on for a while after the wind stops blowing. There is also the aesthetic consideration that wave and tidal power machines, lying on or below the sea surface, are unobtrusive compared to wind turbines. Inhabitants of the Scottish island of Lewis, for instance, voted against a wind farm on their island, but have accepted a project designed by the Wavegen company for one of their bays.

However, ocean power has drawbacks. The biggest is the very power of the sea – the capacity of storms to destroy man's attempts to harness its energy. The designers of wave and tidal technologies have to strike a very difficult balance. On the one hand, they have to design machines that in contrast to ships or offshore oil rigs will in normal weather absorb maximum impact from waves and tides, in order to extract maximum energy from the ocean. On the other hand, they have to ensure that their wave and tidal machines stand up to storms or very rough weather.

This safety priority is a concern with a precedent in wind turbines, which are designed to stop generating during extreme storms to preserve their electronics and machinery (as well as to avoid overloading the grid). Conditions at sea, however, can be even rougher (because of the greater density of water) and designers of wave and tidal power machines have to build them so that the machines absorb less energy as waves get bigger or as currents get stronger. As a result, ocean power development is probably twenty years behind that of wind or solar power. Today there are many theoretical prototypes for tidal or wave machines, but no settled agreement on the best design for manufacturers to focus on.

One reason why ocean power development lags behind other forms of renewables is that it is a minority activity among the nations of the world. Some countries are land-locked, while others are "cursed" with relative calmness in the waters around them. Of the dozen countries actively pursuing ocean energy, the UK is probably the most blessed with ocean energy resources.

Britain is an island sandwiched between the Atlantic Ocean and North Sea. It is largely exposed to the full impact of Atlantic breakers that have had some three thousand miles to be whipped up by the prevailing westerly winds. The North

One of Wavegen's turbine's in situ in Scotland.

Sea is generally less wave-tossed, but completes the enormous tidal pool in which the UK sits. Along the northern tip of the British mainland is the Pentland Firth, a channel through which daily rip tides rush back and forth between the Atlantic and the North Sea. It could be called the Saudi Arabia of ocean energy (if anyone could design equipment robust to withstand its turbulence). It also has, in the Severn estuary, an exceptionally favourable conduit for tidal power, should it choose to exploit it.

Tidal power

Tidal power itself can be divided into two categories: tidal barrages and tidal streams.

Tidal barrages

Some efforts to exploit tidal power with barrages go back many centuries. The eleventh century *Domesday Book* mentions a tidal mill at Woodbridge in Suffolk in the UK. This had gates that let in the incoming tide, but which shut the moment the water started to go out. It effectively dammed up seawater and released it to drive a water wheel that could

The La Rance tidal barrage is the world's biggest.

grind corn. But, despite this historical pedigree, there are today only four sizeable, working tidal barrages in the world, all built in the last fifty years and none of them (yet) in the UK.

The biggest is the barrage that the French built over the Rance river in Brittany in the 1960s. Three smaller ones exist at Kislaya Guba on the Kola peninsula in Russia, at Jiangxia in China and at Annapolis Royal in Canada's Nova Scotia province. Apart from the latter, all of them operate in both directions, generating electricity from both the incoming and outgoing tides passing through the barrage.

The Annapolis Royal barrage, like the ancient Woodbridge one, operates in only one direction, trapping incoming water and releasing it to generate power when the tide has gone out. This Canadian barrage is, in fact, the best placed of all, being situated on an inlet of the huge Bay of Fundy, which has the world's highest tide differential – seventeen metres (between high and low tide). This long (270 km) bay between Nova Scotia and New Brunswick is ideally shaped to produce record high and low tides because, as it extends east, it grows narrower as well as shallower.

The main objection to tidal barrages is the same as to hydroelectric dams on the non-tidal parts of rivers around the world – the ecological impact on animals, plants and soils. These environmental issues will play a big part in the UK debate about whether to build a barrage across the Severn estuary at the top of the Bristol Channel. It has the same combination of diminishing width and depth, and thus produces tides almost as dramatic as the Bay of Fundy. A possible sixteen-kilometre barrage across the Severn could generate as much as five percent of the UK's electricity, at a cost of some £15bn, but also at the price of disrupting rich fish and bird-breeding areas.

Tidal stream

Placing turbines into fast-flowing currents or tidal streams is a far more recent idea. It could also be taken up more quickly, because it does not have the environmental drawbacks of barrages. The Marine Current Turbines company has placed a tidal power turbine in Strangford Lough on the coast of Northern Ireland, and MCT is working with RWE, a German-owned utility company on a project off North Wales.

Another company, Lunar Energy, is planning to install turbines off the Pembrokeshire coast with Eon. And OpenHydro, an Irish company, has one of its tidal turbines hooked up to the UK grid at the European Marine Energy Centre in Orkney.

The Seagen tidal power turbine in Strangford Lough, on Northern Ireland's north coast, photographed above the water and depicted below via an artist's impression demonstrating the turbine's blades beneath.

Owned by Marine Current Turbines Ltd, Seagen is the world's first large-scale commercial tidal energy converter. Installed in April 2008, it was connected to the national grid in July that year, and generates 1.2 megawatts of electricity for between 18 and 20 hours a day.

The turbines have a patented feature by which the rotor blades can be pitched through 180 degrees. This allows it to exploit water flowing in both directions – as the tides come in and out of the straits of the lough.

Outside the UK, there are many potential sites for tidal stream power – the Strait of Gibraltar, the Bosphorus, the Cook Strait in New Zealand, the Bass and Torres Straits in Australia. But don't count on seeing anything from the Mediterranean, where the tide is minimal.

Wave power

Wave power is harder to capture than tidal power, because waves are a more chaotic form of motion than tides or currents. The up, down and rocking motion of waves is more difficult to transform into the rotational motion required for electricity generation.

However, this was achieved in the early 1970s by Professor Stephen Salter of Edinburgh University, with a device that was dubbed Salter's Duck. It was essentially composed of floating containers, tied together and loosely tethered to the sea floor. The waves rocked these containers, which converted the rocking into a rotational motion that could spin a generator. Development of this was dropped when public funding was cut in the 1980s. But the idea has lived on in the work of some of Professor Salter's Edinburgh pupils in a company called Pelamis Wave Power. Pelamis (the name of a sea snake) has installed several of its

A Pelamis Sea Snake wave energy converter, pictured off the coast of Portugal. It uses the motion of ocean surface waves to create electricity, and is made up of connected sections which flex and bend as waves pass.

Durability has always been a problem with wave-power electricity generators, but the Pelamis's design and long, thin shape allows it to resist powerful hydrodynamic forces. Its joints can even be adjusted to match the environment it is installed in: to maximize capture efficiency in calm sea or turn it down for a long life in more violent waters.

machines – each one composed of huge cylinders linked in a sinuous chain – in waters off Portugal to generate wave power for the Portuguese grid (Portuguese feed-in tariffs make this profitable).

There are other designs for wave-power generators. One innovation that would help cut costs is a floating machine called the Wave Treader, produced by Green Ocean Energy of Aberdeen. This undulating device is attached to the fixed base of an offshore wind turbine, with the idea of sharing infrastructure costs with wind operators. Reducing costs is very important for wave and tidal power developers, who need to make their machines sufficiently robust to withstand storms, but not so much as to price themselves out of the market.

For although people may willingly pay extra for a product – be it a Toyota electric hybrid Prius or heather honey – they won't do so readily for a commodity. Max Carcas, Pelamis' business development manager, put it like this in an interview with Greentechmedia: "You don't really invite your friends home, saying 'Do come and see the quality of my electricity. It is produced by artisan wave engineers from the north of Scotland'.

Dams and hydropower

The traditional and very effective method of generating hydropower has been to construct a dam across a river, let the mass and the level of the water build up behind the dam, and then let the water fall at speed into turbines at the bottom of the dam. So successful has this been that this kind of hydropower generates no less than sixteen percent of the world's electricity – far more than any newer source of renewable energy. Hydropower has none of the short-term variability affecting most other renewable energy sources – though of course it is vulnerable to drought.

But dams – big dams, at least – have fallen out of fashion. Anti-dam campaigners, ranging from the US-based International Rivers Network to the award-winning Indian novelist Arundhati Roy, have highlighted the problems of dams in terms of population displacement, loss of river species and even increases in greenhouse gases in tropical countries. In hot climates the rotting vegetation which can build up around dams produces a lot of methane and carbon dioxide. Partly as a result of NGO pressure, a World Commission on Dams was set up in 1997, and in its widely praised report of 2000, it acknowledged that "coercion and violence have been used against communities affected by dams", that dam projects had to win "greater public acceptance" and that dam-project assessments should also consider other ways to provide the services of energy, flood control and irrigation offered by dams. The World Bank is now much more cautious about financing new dams in developing countries. In the US, for environmental reasons, the rate of decommissioning existing dams now exceeds the rate of commissioning new ones.

But the age of dams (or "dam-age", as NGO critics call it), is certainly not over. China has just built the Three Gorges Dam on the Yangtze river, the building of which has required the displacement of more than one million people. This is the world's largest dam in terms of generating capacity (some 25,000 MW), relegating the Itaipu dam on the Brazil-Paraguay border (which has a capacity of 14,000 MW) to second place.

Much less controversial are the "run-of-the-river" hydropower projects. These do not involve any damming or creation of reservoirs – just some walls or conduits to channel part of the river water into turbines. They are usually small projects. But one such project under discussion in Africa is enormous. The two Inga dams currently operate at a low-output level, despite the fact that they are situated at a point where water flow is huge and the level drops nearly one hundred metres on the Congo river in central Africa. The plan is to rehabilitate them for the construction of

The old Inga dam, on the Congo river, is to be radically expanded and updated, to rival, and possibly even surpass, the electricity output of China's Three Gorges.

two new dams, the Grand Inga and Inga III. If the project is realized, the dams could produce more than Three Gorges, and as much as one third of Africa's entire electricity output.

Backers of the project say it would "light up" the dark heart of Africa. Critics say it would do nothing of the kind, because most central Africans are not on the grid, and much of the output would be lost in transmission to distant customers.

Biomass and biofuels

The power of vegetation

In all its variants, of which biofuel is only one, biomass is by far the most important renewable source of energy: it accounts for around ten percent of global energy supply. But to an extent this figure has to be an educated guess, because much of this is material of a very basic nature – farm waste, dung, wood and wood residue – that never enters the commercial sector and therefore never reaches official statistics.

Most of this is produced and utilized in developing countries, especially south Asia and sub-Saharan Africa, and particularly places that aren't connected to an electricity grid. Indeed, Africa accounts for only 5 percent of the world's energy supply, but 25 percent of the world's solid biomass used for energy. While solid biomass is the largest renewable source, it is also the slowest-growing renewable sector, increasing at 1.5 percent a year on average since 1990.

Concerns about energy security and climate change have encouraged some greater use of solid biomass in industrialized countries as well. In the power sector, it is increasingly being burned alongside coal in a process known as "co-firing" – essentially the burning of two different fuels in the same combustion device. A maximum of only ten percent of biomass content can be added without the coal plant having to be re-configured. It's a long way from being green, but every chance to burn something cleaner than coal should be taken.

However, the real bio-energy revolution lies not in the burning of solid biomass, but in the controversial use of biomass in liquid form – as biofuels for road transport.

Different generations of biofuels

First generation

First generation biofuels are processed from crops. They have a lot in common with that well-known biofuel, alcohol. There are two main types. Bio-ethanol, which is made in large quantities in Brazil and in the US, is processed by taking food crops such as corn maize, releasing the sugar and fermenting it. It has advantages over ordinary petrol or gasoline, because it has a higher octane ratio (a fuel's resistance to detonation), which improves controlled burning of fuel, enabling increased compression of air in the cylinder and more efficient combustion.

Apart from having – like all substitutes for fossil fuels – a lower energy density, ethanol's main disadvantage is that it is corrosive. Most car engines

Construction underway at the main processing works for one of the largest biomass co-firing projects in the world – the 4000 MW power station in North Yorkshire. Owned by Drax Power Ltd, it will co-fire renewable biomass materials alongside coal in its boilers. The company has set itself the target of producing ten percent of its output from co-firing.

Iogen operates the world's first demonstration facility of second-generation biofuels. Opened in 2004, it makes clean-burning cellulosic ethanol fuel from agricultural residues. Wheat, oat and barley straw are used as raw materials.

can take a mix of up to fifteen percent ethanol without modification; higher concentrations, however, require adaptation. Bio-diesel usually uses a different set of crops, notably oil seeds such as soy and rapeseed. It is an oxygenated fuel, meaning that it is cleaner and improves combustion, compared to ordinary diesel. And conventional diesel engines do not need much modification to run on even one hundred percent bio-diesel.

How green are biofuels?

Confusingly, the strongest arguments for and against biofuels are both environmental, and are specifically related to climate change.

Biofuels still emit some CO_2, albeit less than oil. So how do they help fight climate change? The answer is that they are at least in theory carbon-neutral – in the sense that when burned they merely return to the atmosphere the carbon that they absorbed when they were growing plants.

Granted, you could say the same about oil – that oil, when burned up, just returns to the atmosphere the carbon it absorbed from plants and algae millions of years ago. But the difference is that we can't repeat oil's absorption of carbon, which took place under conditions of high temperature and pressure that can't be replicated, whereas we can repeat biomass's absorption of carbon by planting trees or energy crops – which is what make biofuels renewable.

Second generation

Second-generation biofuels process fuel from non-food biomass and are not yet commercially available. They derive from cellulosic substances such as wood and grassy products and by-products, rather than food or feed crops needed for people or livestock.

But extracting sugars from them is not easy. All plants contain cellulose and lignin – two complex carbohydrates which are tough to break down. Cattle manage to digest cellulose and turn it into glucose, but they have a very complex digestive system with which to do it. Which, incidentally, means it is hardly surprising that cows belch so much: most of the methane emissions of cows derives from their belching (and not, as is commonly believed, from farting).

Third generation

Third-generation biofuels are derived from algae. The idea here is to make biofuel with algae – or "oilgae", as they is dubbed. Algae, essentially slimy underwater plants, are grown in water infused with CO_2. The algae

In 2007, a team of University of Georgia researchers announced that they had developed a new biofuel derived from wood chips. Unlike previous fuels derived from wood, the fuel can be blended with bio-diesel and petroleum diesel to power conventional engines.

A paddlewheel circulates the water in one of New Mexico's many algae cultivation ponds.

convert the CO_2 into oxygen and biomass during photosynthesis and the biomass can then be turned into biofuel.

Algaculture need not affect fresh water resources, as algae can be produced using ocean and wastewater. Its by-products are biodegradable and, compared to other energy sources, are relatively harmless to the environment if spilled. As part of the American Recovery and Reinvestment Act, the US Department of Energy (DOE) announced in July 2009 that it would be spending $85 million on developing algae-based biofuels and advanced biofuels that would be compatible with the existing infrastructure.

Biofuels and transport

Both bio-ethanol (made mainly from sugar cane or maize) and bio-diesel (made generally from oil seeds such as rapeseeds and soybeans) could theoretically be useful in decarbonizing our transport sector. This is a particularly difficult task: around 98 percent of road transport in the UK, and in most countries (with the exception of Brazil – see box on p.143), is fossil-fuelled. Most of the future growth in CO_2 emissions is expected to come from transport. And biofuels are the only cleaner alternative road-transport fuel on the near horizon (see separate sections on electric cars and hydrogen).

Among the political and economic reasons for favouring biofuels are that they exert a downward pressure on the oil price and cut countries' oil import bills; if grown at home, they also increase a country's energy security. Moreover, far more countries have the potential to grow and perhaps export biofuels than to produce oil or gas, the reserves of which are more

The Algeus is the world's first algae-fuelled hybrid car. Like the Prius, it is part-electric, but unlike the Prius it burns biofuel rather than conventional petrol.

restricted geographically. Individuals as well as organizations can produce biofuels, which, unlike oil, can generate income for thousands of farmers – as opposed to just governments and a few oil companies.

Good crop, bad crop

But biofuels are a long way from unproblematic. There are good and bad biofuels, depending on the type of plant or crop, how it is grown and processed, and where it is grown. The whole point about biofuels is that they ought to be a way of cutting down on fossil fuels. Any saving will be undermined if a lot of fossil fuel has gone into the use of fertilizers, tractors and machinery to cultivate and manufacture the biofuel in question. The analysis of the full life-cycle of biofuels – calculating how much fossil fuels goes into the growing and making of biofuels – is known as a "well-to-wheels" analysis. The terminology deliberately references that of the oil industry in order to remind us that, despite the fact that biofuels do not come out of wells like oil, the two oils are comparable.

In terms of a well-to-wheels analysis and its carbon footprint, one of the worst biofuels is ethanol made from maize, as grown in large amounts in the US Mid-West. The maize requires considerable cultivation and does not convert very easily to the sugar needed for it to ferment into the alcohol required for ethanol. By contrast, one of the best biofuels in terms of greenhouse-gas savings is commonly held to be the ethanol that Brazil makes from its sugar cane. It requires relatively little cultivation and is, of course, already a sugar.

But changes in land use are also an issue with biofuels. For there are some types of land – wetlands, grasslands, forests and above all tropical forests – which store enormous amounts of carbon in their soil. The conversion of these types of carbon-rich soil to biofuel cultivation might release such large amounts of carbon that no biofuel "saving" could ever make up the carbon loss from the original land-use change – or at best only make up the loss over a period of many years.

This issue has raised doubts about how good a biofuel Brazilian ethanol actually is. Brazil insists that it grows its sugar cane far from the Amazon rainforest region. However, critics complain that expansion of sugarcane growing in the south pushes other activities, such as cattle-rearing, further north and adds pressure to clear the Amazon basin for cattle ranches, thus encouraging the cutting-down of jungle and forests that absorb carbon.

This displacement effect – of fuel crops displacing food cultivation, which then moves elsewhere – is the main environmental objection of levelled against biofuels by NGOs and others. A year after the UK introduced its Renewable Transport Fuels Obligation (RTFO) in 2008, Friends of the Earth complained that in its calculations, the government ignores "how much forest is being cut down to replace food crops that have been displaced in order to grow biofuels for the UK".

Fuel versus food

The way in which fuel crops may be displacing food crops and leading to additional deforestation is part of the wider argument over whether biofuels are overstraining food and indeed water resources. This argument came to the fore as food prices, along with other commodity prices, came to a peak in the first part of 2008. The US administration and the European Commission, which had both by that time committed themselves to ambitious biofuels targets, admitted biofuels had pushed food prices up, but were only responsible for a mere three percent of the increase. However, a 2008 study commissioned although not officially endorsed by the World Bank blamed biofuels for 75 percent of the 140 percent rise in the price of a given basket of commodities over the period 2002–08.

Sharper criticism came from Jean Ziegler, the United Nations Special Rapporteur on the Right to Food between 2000 and 2008. He called crop-based biofuels "a crime against humanity". Typical of the strong opposition by many NGOs to biofuels was the Oxfam report in 2008, entitled "Another Inconvenient Truth" (echoing the title of Al Gore's book on climate change).

This argued that biofuels were solving neither the climate crisis nor the fuel crisis, but instead were contributing to food insecurity and inflation. "If the fuel value of a crop exceeds its food value, then it will be used for fuel instead", the report found. "Thanks to generous subsidies and tax breaks, that is exactly what is happening … rich countries spent $15bn last year supporting biofuels while blocking cheaper Brazilian ethanol, which is far less damaging for global food security", the Oxfam study said.

The tension between fuel and food would disappear if second-generation biofuels could be properly exploited. As a result, research into them is being stepped up, both in terms of public money and regulation. The European Union has passed a renewables directive, requiring each of its 27 member states to ensure by 2020 that 10 percent of all its transport fuel consumption

Jean Ziegler in 2009 at the Auditorium Maximum of the University of Vienna, which was occupied by protesting students. He has dismissed crop-based biofuels as an "affront to humanity".

should be renewable. Under this law, second-generation biofuels will get a double credit towards the ten percent target.

Trade and standards

Biofuels can bring economic and social advantages and, with care, environmental benefits. At present the trade in biofuels is relatively small. The amount of biofuels that is traded across national boundaries constitutes only around ten percent of all biofuels produced or consumed (whereas nearly fifty percent of all oil in the world is traded from one country to another).

Yet many developing countries would have a comparative advantage in producing biofuels for export. Furthermore, it would be environmentally very damaging for, say, Europe or the US to do all the substitution for oil with home-grown biofuels without regard to the comparative advantage of countries like Brazil. The European Union's Biofuel Research Advisory Council estimated in 2006 that "in 2030 EU biomass would hold the

Brazil's ethanol economy

Brazil is the world's number two producer of ethanol (behind the US) and the number one exporter of it. As the result of a thirty-year government promotion programme, slightly over half its huge sugarcane harvest now goes into ethanol production. In addition to efficiencies of scale, Brazil's manufacture of ethanol from sugar cane benefits from use of bagasse, the fibrous residue of cane after the sugar has been squeezed out of it, to heat and power the fermentation of ethanol.

Using bagasse, most ethanol plants are self-sufficient in energy and can even sell some excess electricity to the grid. It is reckoned to be competitive with oil if the oil price is at least $30 per barrel. And it is more than competitive with US corn-based ethanol (chiefly for the extra initial processing that US manufacturers must carry out – turning their corn into sugar). The only way US corn-ethanol producers keep Brazilian sugarcane-ethanol producers from overrunning the US market is by persuading Washington to put a hefty import tariff on ethanol.

Despite the intrinsic competitive edge of Brazilian ethanol, it has also had considerable government support. The state has encouraged the supply side of the industry through fixed purchases of ethanol by Petrobras, the state oil company, low-interest loans to ethanol refiners and price support for ethanol. The state has also mandated demand for ethanol. Since the 1970s, blending ethanol with petrol has been required, and currently the minimum ratio of ethanol is 25 percent (though this has sometimes varied depending on the level of sugarcane harvests).

The car industry in Brazil, which includes most of the European, Japanese and US car majors, makes "flexi-fuel" cars able to run on proportions of ethanol that range from 25 percent up to 100 percent. About a quarter of the country's total car fleet is now flexi-fuel, while the share of flexi-cars in new sales every year is now around 90 percent. Brazil still imports oil for diesel trucks, although these imports may decrease in view of the major new oil finds in the Brazilian offshore. Adding in all the country's diesel vehicles, ethanol now accounts for around seventeen percent of total road-fuel consumption.

technical potential to cover between 27 and 48 percent of our road transport fuels needs, if all biomass would be dedicated to biofuel production". While this could be seen as an encouraging statistic, it is nonsensical to speak of dedicating all biomass to biofuel production. So the advisory council suggested that covering a quarter share of EU road-transport fuel needs by 2030 would be realistic – half from domestic production and half from imports. So developing countries deserve a market for their biofuel exports. At the same time, it is clear their exports will have to meet some basic environmental standards.

Hydrogen

The fuel fervour that faded

The hydrogen story is a salutary reminder of how alternative energies can come in and out of fashion, especially if they are over-hyped. In his 2003 State of the Union address, president George W. Bush proposed funding so that "the first car driven by a child born today could be powered by hydrogen and pollution-free". He went on to oversee more than $1bn spent on hydrogen fuel-cell research and development during his two-term presidency.

In 2009 Steven Chu, President Barack Obama's Nobel Prize-winning energy secretary, abandoned funding for hydrogen fuel cells in cars. "We asked ourselves", said Mr Chu, "if it is likely in 10, 15, 20 years that we will convert into a hydrogen car economy. The answer we felt was no". Chu did state that hydrogen fuel cell research would be maintained in stationary applications such as back-up for power stations.

The decision was denounced by the US National Hydrogen Association. It complained that the Obama administration was throwing away the fruits of past research and protested that hydrogen fuel cell vehicles were "not a science experiment, but real vehicles with real marketability and real benefits". But by 2009 a number of big car companies were already seeing more marketability and benefit in electric cars. (The ditching of research into hydrogen cars in favour of electric cars and hybrids will doubtlessly produce a conspiracy theory, in the same vein as the theory that the hydrogen lobby was one of many groups who conspired to get General Motors to pull its electric car, EV1, in the 1990s.)

Hydrogen is a very appealing prospect on paper. It is the simplest element, having only one proton, and is the most universal gas in the universe. It also has the highest energy content of any common fuel by weight – three times more than petrol – because it is so light. But it is found only in compound form with other elements, from which it must be separated (either chemically or by electrolysis) to be useable. As a carrier of energy that must be produced from another substance, hydrogen is similar to electricity, but unlike electricity, it is storable. Hydrogen fuel cells combine stored hydrogen with oxygen to create electricity, and the only by-product is pure water vapour. (So pure is this water vapour that on NASA space shuttles – themselves powered by liquid hydrogen rockets – the crew drink the condensed water vapour from hydrogen fuel cells used to make onboard electricity). So hydrogen can produce electricity without any direct CO_2 emissions. This does, however, leave the issue of indirect emissions: hydrogen does after all require an energy source to make the hydrogen in the first place.

In 1874, the science fiction writer Jules Verne suggested that "water will one day be used as a fuel" in his novel *The Mysterious Island*, and

Iceland has made the biggest commitment to hydrogen of any nation, aiming to transform to a "hydrogen economy" by 2050. As part of the ECTOS demonstration project, which ran from 2001 until 2005, three hydrogen fuel cell buses and one fuel station were built. All of the energy used to produce the hydrogen derived from Iceland's renewable energies, making their hydrogen use wholly carbon-free.

The fate of the enormous *Hindenberg* passenger zeppelin did not inspire confidence in hydrogen-powered transportation.

many engineers in the early part of the 20th century were excited by its possibilities. It received a very bad setback in the 1937 Hindenberg disaster, when the hydrogen-filled German zeppelin burned up, killing many of its passengers. Interest in it revived in the 1970s with the rise in oil prices and the susearch for alternative fuels. Pro-hydrogen fuel enthusiasts have waxed lyrical about "the hydrogen economy", the scale of which is partly due to hydrogen's abundance, but also a recognition that its use as a fuel would require an infrastructure revolution. (One writer, Jeremy Rifkin, wrote a book with the wonderfully understated title of *The Hydrogen Economy: The Creation of the Worldwide Energy Web and the Redistribution of Energy on Earth.*)

The problems of storage and infrastructure that would have to be solved for a hydrogen economy to emerge have become more and more evident over the years. Hydrogen has to be stored at pressure, and requires special pipelines or containers for transport. (Some people thought in the 1970s that the natural gas pipeline network could be used, with an increasing ratio of hydrogen fed into it – an impractical idea for many reasons). Talk of the hydrogen economy seems to be partly a celebration of hydrogen's abundance, but it's also partly a recognition that its use as a fuel would require an infrastructure revolution. As a result, a classic chicken-and-egg dilemma has arisen: the car companies do not want to make hydrogen fuel-cell cars until there is a network of garages stocking hydrogen, and the oil companies do not want to build this network of hydrogen stations until there are cars to use it. There have been discussions between car and oil companies to break this impasse.

Furthermore, electric batteries have enormously improved, thereby reducing hydrogen's storability advantage over electricity. Hydrogen cars are essentially just electric cars with their electric battery substituted by a fuel cell and a little storage tank of hydrogen. If that hydrogen is going to contribute to mitigating climate change, it will have to be renewably produced. And that energy source will almost certainly be renewable electricity. Is it not, therefore, rather a detour to use renewable electricity to produce hydrogen in order to get renewable electricity out of a hydrogen fuel cell – particularly when a substantial amount of renewable electricity could be put straight into a car battery and reliably stored within it? Moreover, the progress in on-board electric battery storage for cars comes

A bus is refilled with hydrogen fuel in China's first hydrogen station, located in Beijing Hydrogen Park. Jointly built by BP and its Chinese partner Sinohytec, the station provides 100 kilograms of hydrogen a day, enough to fuel four hydrogen-powered buses.

in addition to an electric grid (for home or mid-journey plug-in refills) that is a far more extensive energy network than hydrogen could ever be.

In 2000 Bill Ford, chairman and CEO of his family car company, famously said "I believe that fuel cells will finally end the 100-year reign of the internal combustion engine". But by 2008 the Ford company was announcing that in product development "its next step is to increase over time the volume of electrified vehicles, as the transition from hybrid to plug-in hybrid to battery electric vehicles occurs". Other companies were making the same shift of emphasis from hydrogen to electric cars.

None of this, however, means the general abandonment of hydrogen fuel-cell development. Where compact storage is not a priority – as in houses or factories – fuel cells can play an important role. They may even perhaps have some mobile uses in, for instance, powering urban bus fleets which could operate from a centralized hydrogen distribution point. For this reason, many cities – London included – are still continuing to maintain and run hydrogen buses.

PART 3
THE
PLAYERS

The people and
companies in
control

From the Seven Sisters to the NOCs

Where the big energy companies came from

The original "seven sisters" were the biggest companies in their countries, and they dominated the world oil scene of their day. The phrase was coined around 1950 by Enrico Mattei, founder of ENI, Italy's oil major, in irritation at the seven predominantly Anglo-American companies, which seemed to have all the international oil concessions locked up. Over time, of course, the composition of the septet has changed.

The seven sisters Mattei had in mind were Standard Oil of New Jersey (still with us as Exxon), Royal Dutch Shell, the Anglo-Iranian Oil company (now BP), Texaco, Socony-Mobil, Gulf and Socal. Since then Mobil has merged with Exxon, while Gulf, Socal and Texaco have merged to create Chevron. Chevron would be in any updated list of the top seven international oil companies, as would Total of France, Mattei's own ENI, and ConocoPhillips of the US.

These are still enormous companies. Exxon has the biggest stock-market capitalization in the US, and it's often referred to, alongside most of the other top six, as a "super-major". The super-majors grew even bigger in the late 1990s in a wave of mergers. But these were partly defensive mergers by companies which had come to realize that, however big they might seem within their own Western industrialized economies, they were being dwarfed by the state-controlled oil companies of the major oil-producing countries, mostly members of OPEC but also including Russia, Mexico and Brazil. In terms of reserves and even production, the international oil companies now lag far behind national oil companies (NOCs) like Saudi Aramco, the National Iranian Oil Company, PdVsa of Venezuela, Petrobras of Brazil or Gazprom of Russia.

This is no surprise. These companies and their countries hold most of the world's oil

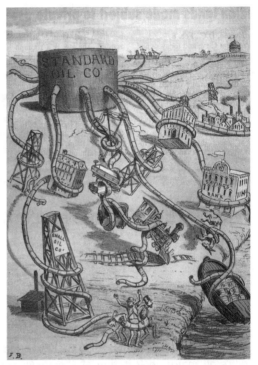

Public concern over the power of oil companies goes back a long way, as demonstrated by this political cartoon from the nineteenth century, pointing out just how powerful Rockefeller's Standard Oil company had become in the US.

reserves. These reserves were first developed, for the most part, by the international oil companies, or IOCs as the Western publicly listed and traded oil companies are referred to in the oil business. But over a period from the 1950s until the mid-1970s, the oil-producing countries asserted sovereignty over their reserves and passed control of these reserves to their own newly-created national oil companies, or NOCs as they are referred to in the oil world. This, plus the fact that non-OPEC oil exploration success in areas like the UK part of the North Sea and the US portion of the Gulf of Mexico has peaked, means that a little over seventy percent of world oil and gas reserves is now in the hands of NOCs, though their share of global oil and gas output is just over fifty percent.

This shift of power in the oil industry is part of the past century's transformation from colonialism to globalization. Oil began as a home-grown industry in the US, led by American entrepreneurs and companies such as the Rockefellers and their Standard Oil company, which eventually sought foreign concessions to expand overseas.

But from the start the European oil business was mostly a matter of obtaining concessions from developing countries. This was how the Anglo-Persian Oil Company (the precursor of BP) got started in Iran, and the Compagnie Française des Petroles (a forerunner of Total) began in Iraq. The start of Royal Dutch Petroleum in Indonesia was more overtly colonial because Indonesia was a Dutch colony, part of the Dutch East Indies at the time.

Concessions

In returning for paying an upfront fee and subsequent tax or royalties on production, a concession generally allowed a company the exclusive right to explore for oil and to produce oil within a given area for a set period of time. It usually left the company free to set exploration programmes, production levels and prices.

This system gave the IOCs considerable autonomy, and left host governments with little control over, or even knowledge about, the day-to-day management of their principal resource. The concession system was eventually swept away in the mid-1970s in all the major oil-producing countries. Long before then, however, some governments had begun to take action.

Fifty-fifty

The first instances of resource nationalism came in Latin America, which is not surprising given the region's long history of invasion, colonialism and anti-colonialism, going back to the early nineteenth century liberation from Spain. In 1938 Mexico expropriated (with compensation) all foreign oil company assets and operations, and created Pemex to run them. In 1942 Venezuela decided to renegotiate all its concessions, and began a series of tax increases so that by 1947, the total "tax take" of royalty and tax amounted to fifty percent of all profits. By 1950 a similar fifty-fifty profit split was agreed in Saudi Arabia, where oil had been found in 1938, between US companies and the government.

Iran wanted to achieve the same kind of profit-share with Britain's Anglo-Iranian Oil Company. But Britain resisted, in a way that still colours Anglo-Iranian relations today. Acutely aware of its oil dependence on Iran, the UK government had already shown a tendency to treat Iran as a colony: indeed UK troops occupied part of it during World War II. So in 1951 the Iranian government of prime minister Mohammed Mossadeq decided to nationalize Anglo-Iranian. In 1953 Britain and the US supported a coup against Mossadeq and restored the Pahlevi dynasty of the Shahs. Britain got back part of its concession, but had to share the rest with the US.

Not all of Mossadeq's work was undone. The National Iranian Oil Company that he set up remained in being as the owner of the oil, with the status of foreign oil companies reduced to that of sub-contractors to NIOC.

Nationalizations and nationalism

The year 1971 brought a bumper crop of oil nationalizations. There were many reasons for other oil-producing governments to finally follow the lead set long before by Mexico and Iran. Among them were a tighter oil market giving producers more leverage; a steady decline in the standing of the US in world opinion because of Vietnam, and of Britain and France in Middle East opinion since their disastrous joint venture at Suez in 1956 (both of which made retaliation less likely); and frustration with the power of the US and European concessionaires.

Most believe that 1973 was the first time that the Arab oil producers wielded their "oil weapon", with their boycott of the US for its support of Israel during the October 1973 Arab-Israeli conflict. In fact, in response to the 1967 Arab-Israeli Six Day War, the governments of Saudi Arabia, Kuwait, Iraq, Libya and Algeria banned shipments of oil to the US and UK for backing Israel. But the move had almost no impact at all, because the IOCs ran the international oil trade and asked Venezuela and Iran which was not only not an Arab country, but at that time friendly with Israel to pump more oil.

In that year, Algeria completed its takeover of foreign oil assets by nationalizing French oil interests, Iraq pursued further its nationalization of foreign stakes in the Iraq Petroleum Corporation, and Libya's new Colonel Moammar Qaddafi nationalized BP assets. This last act was a purely political move by Qaddafi, who took over BP in protest at Britain's

Bolivian army soldiers monitor the premises of an oil refinery, following the nationalization of Bolivia's gas and oil industry. In May 2006, Bolivia's President Evo Morales decreed that all foreign energy firms must agree to channel all their sales through the Bolivian state or else leave the country, and gave a six-month deadline.

The main foreign oil firms operating in Bolivia then were Brazil's Petrobras, the Spanish-Argentine company Repsol YPF (shown here), British companies British Gas and BP, France's Total, and the US Exxon Mobil Corporation.

failure to stop Iran taking control of two small islands in the Strait of Hormuz, when Britain withdrew all its forces from the Gulf in 1971. Arguably, BP which was progressively privatized from the mid-1970s to the mid-1990s suffered more than it gained from UK government ownership.

Taking over local BP assets became a convenient way of showing displeasure with London. Another example of this came in 1979 when the Nigerian government nationalised BP's assets in Nigeria, in order to put pressure on Britain to take a tougher line in dealing with the breakaway white government of Rhodesia. Meanwhile, Shell, with its British identity diluted by being in harness with Royal Dutch Petroleum, was allowed to stay on in Nigeria, as have many other IOCs.

The era of nationalizing foreign oil company assets is by no means over, especially when a rising oil price increases the desire of host gov-

Russian Prime Minister Vladimir Putin at the opening ceremony of the Gas Exporting Countries Forum (GECF) in Moscow, December 2008.

ernments to increase their share of the oil revenue, regardless of contractual niceties. Two recent cases of resource nationalism are Venezuela and Russia, both countries whose current leaders feel that their predecessors gave away too much to foreign oil interests in the 1990s.

In 2006 Venezuela's President Hugo Chavez announced that he was effectively nationalizing all oil fields by raising the state company PdVsa's stake in all operations from forty to sixty percent. The fact that the left-wing president made this move seven years after coming to power might reflect a certain initial caution, or more probably that after a steady run-up in the oil price over that period, he could no longer resist the financial temptation to get a bigger slice of rising oil revenues.

The move was accepted by most foreign companies (in return for compensation). But ExxonMobil, with characteristically robust insistence on contract observance, refused to countenance it, leading to the seizure of its assets by the Chavez government. To put pressure on Venezuela for compensation, ExxonMobil has tried to get courts in the US and Europe to freeze PdVsa's foreign assets. Chavez has threatened to cut off all oil shipments to the US should that happen.

In Russia, Vladimir Putin has made it his business, first as president (2000–08) and then as prime minister (since 2008), to restore some state control over an oil sector that had been divided up and privatized in the 1990s (in sharp contrast to a gas industry that was kept intact and handed to Gazprom). This involved reinforcing the state-controlled Rosneft oil company, and passing legislation to prevent foreign oil companies taking majority stakes in any "strategic" oil and gas fields. Putin took particular aim at what he called the "colonial" nature of production-sharing agreements (PSAs) that a few Western majors had negotiated in the 1990s.

PSAs, which allow an oil company to recover all its costs from oil sales before it has to pay any tax or royalty, act as a kind of financial ring-fencing to protect a project from arbitrary interference in countries judged unstable. And as such they have frequently been used in developing countries – hence Putin's use of the word colonial. Two PSAs that were signed in the 1990s concerned two huge gas projects at Sakhalin island on Russia's Pacific coast, from one of which Shell was effectively pushed out to make room for Gazprom.

These moves owe something to the individual temperaments of Chavez and Putin. But they came against a background of an almost steadily rising oil price from 1999 to 2008, which understandably whetted the appetite of host governments for a bigger slice of oil revenue, regardless of the niceties of contract with foreign partners. So, during the 2000s, other governments revised contracts to get better terms from foreign oil companies as the oil price went up, though none went as far as Venezuela's across-the-board nationalization.

Ironically, if there was one oil-producing country that could not afford, in terms of its own public opinion, to appear a soft touch for foreign oil companies, it was Iraq. Aware of the widespread belief which was not completely unfounded that the US invaded Iraq in 2003 for its oil, the Iraqi government knew it had to take a tough negotiating stand. And so it did, when in 2009 Baghdad started to auction oil leases to foreign companies. Some countries are not as possessive of the exploitation of their natural gas as they are of oil. Saudi Arabia, for instance, would not dream of inviting foreign companies in to prospect, on their account, for oil, but has been happy to let them look for gas.

NOCs – shaped by history

The individual character of NOCs is somewhat shaped by the nature of their transformation from IOC operations. In the case of Saudi Arabia, there was a smooth transition of Aramco (the Arabian American Oil Company) into Saudi Aramco. The Saudis retained the expertise of the US companies that made up the Aramco consortium until they the Saudis felt able to take over. Today Saudi Aramco is rated as the top NOC, not

Chavez's petro-aid

Since 2005 Venezuela has been supplying its oil on preferential terms to fourteen of its Caribbean and central American neighbouring states under a scheme called Petrocaribe. This allows participants to buy a portion of their Venezuelan oil purchase on credit with very soft terms (a 25-year maturity at 1 percent interest per annum). The proportion of oil payable with this credit varies according to the oil price on a sliding scale – when the price is $100 a barrel or more, the oil purchaser can get 50 percent of the oil on the soft credit, but only 5 percent of the oil on credit if the oil price is $15 a barrel. Under the Petrocaribe scheme, Venezuelan oil can also be paid for with local commodities such as bananas, rice or sugar.

The scheme expands to the wider region the sort of preferential oil arrangements that Venezuela's left-wing president has already extended to Cuba (which in return has sent doctors and teachers to Venezuela). There is no doubt the Petrocaribe scheme has been appreciated by Venezuela's smaller and poorer neighbours, especially during the 2004–08 climb in the oil price. On the other hand, provision of cheap oil obviously does not encourage these countries to restructure their energy supply and prepare for a day when cheap oil is no longer available from a post-Chavez Venezuela.

Venezuela had received some advice on energy efficiency, recycling and emission control from London, under an arrangement reached by the ex-mayor, the left-wing Ken Livingstone, in which London received some Venezuelan oil at a discount. The discount was passed on in the form of cheaper bus fares for poorer Londoners. But Livingstone's Conservative successor, Boris Johnson, cancelled the deal on taking office in 2008. Chavez has also provided Boston and New York with cheap oil.

only because its production is by far the biggest, but also because of its professionalism. Similarly, Abu Dhabi sought no confrontation with the IOCs when it came to create the Abu Dhabi National Oil Company (Adnoc), and still allows IOCs to operate on its territory. By contrast, the NOCs of Iran and Algeria bear the hallmarks of their countries' more revolutionary tradition.

But what all NOCs have in common is the wider political, social and economic objectives that their state owners give them. This responsibility is inevitable: after all they are the guardians of their nation's oil, for the nation's benefit. However, it does mean that NOCs' revenues are often diverted to goals other than oil. A dramatic example of this is the use by President Hugo Chavez of Venezuela's oil revenue. In 1997, the year before Chavez came to power, Petroleos de Venezuela (PdVsa) spent $77m on social and cultural activities, but in 2005 it spent $7bn on these causes.

Another example of oil being used for political purposes is Venezuela's Petrocaribe scheme (see box).

It is not unknown for IOCs to spend some of their revenue on social projects, such as the hospitals and schools that Shell has built in the Niger delta. But this is a special case. Here Shell is trying to compensate for the failings of the Nigerian state and to give the people of the oil-bearing delta some sense that they are getting a return on the delta's oil wealth. Shell's self-interest is to reduce local people's temptation to sabotage oil facilities and steal oil production. In general, these sort of social programmes are something one might expect an NOC to carry out, not an IOC.

As a result many NOCs appear to lack incentives to maximize profit, because much of this profit disappears into the state budget for non-oil purposes or into subsidizing the price of oil and oil products.

Can NOCs and IOCs collaborate?

Generally, IOCs would love to collaborate more with the NOCs. The Western majors may complain, often with cause, about the bureaucracy

Russian Prime Minister Vladimir Putin and Venezuelan President Hugo Chavez in Caracas, in April 2010. Putin arrived in Caracas to bolster energy and defence ties with Chavez and launch a $20 billion joint venture to tap the Orinoco heavy oil belt.

and inefficiency of some NOCs and the political interference in their operations. But they know that the NOCs are the big holders of remaining resources. The question is more whether the NOCs will allow collaboration. NOCs or their governments may well harbour some residual suspicion of the IOCs for always wanting to claim ownership of reserves wherever possible (so that they, the IOCs, can "book" the reserves onto their balance sheets in order to reassure investors that they have a future).

Moreover, NOCs tend to take a more cautious approach to depletion (the rate at which oil or gas is extracted), whereas IOCs almost always want to pump oil or gas as fast as possible without damaging the oil or gas reservoirs in order to get the payback on their investment. But in fact, the desire of NOCs and their governments to work with the IOCs goes up and down with the oil price. When the oil price is low, and cash is short, oil-producing governments are often happy to have the IOCs investing in their fields. But they have much less need of help from IOCs when the oil price is high and the living is good.

The industry workhorses

Unlike the integrated oil majors whose filling stations make them household names, oil services companies have no public image. Yet they do much of the work – providing seismic services to help the oil companies decide where to drill, doing the drilling, and then maintaining and repairing oil wells. Their role has increased in recent years. The international oil companies (IOCs) have tended to hand over to service companies more of the middle stages of the oil cycle between exploration and refining and marketing. This has enabled the IOCs to limit the number of full-time employees they have, and put more of the burden of adjusting to oil industry booms and busts on to the service companies.

So oil services tend to be the most up-and-down part of a cyclical industry. They tend to be the first to feel the booms and busts. Many people in the sector, and the rigs they use, are paid and hired on a day rate, and can be laid off at very short notice. For instance, although it still has 77,000 employees of 140 nationalities working in 80 countries, Schlumberger laid off 12,000 employees between autumn 2008, as the oil price plummeted, and summer 2009, as the oil price recovered.

Meanwhile the service companies are not only important, but also attractive, to the national oil companies (NOCs). In contrast to the IOCs, they don't try to book oil field reserves as their own, or to try to get the

Is there a future for IOCs?

▶ **Are the IOCs obsolete?** This is a reasonable question. For in recent years there has been a technical "hollowing out" of some IOCs, which increasingly rely upon oil-service subcontractors in order to cut staff overhead costs and increase labour flexibility. BP is a good example. It used to have a big scientific research division and considerable in-house engineering capability. The science laboratories have gone, and BP now contracts out a very large amount of its engineering to the oil service companies. Some would argue that the industrial accidents and leaks at BP operations in Texas and Alaska in the last decade are the price BP has paid for such labour flexibility and outsourcing. Shell has gone some of the way down the same road as BP, though Exxon still retains capabilities across the oil field spectrum. Meanwhile, the oil service companies, especially the two big ones of Halliburton and Schlumberger, have steadily expanded, to the point that they can satisfy all the technical needs of the NOCs.

▶ **Might the oil service companies supplant the IOCs?** No. They already have, in terms of their work with NOCs in developing countries. Yet the oil service companies don't seem to want go head to head in competition with the IOC oil majors. "I will not bid against our customers among the IOCs", the chief executive of a major oil service provider has protested, explaining why he and other service providers leave it to oil companies proper – whether IOC or NOC – to bid for main contracts or leases of oil and gas fields. "I won't take reservoir or pollution risk [the risk of not finding enough oil or accidentally spilling too much] because I don't have the balance sheet to do so."

▶ **So is there still a role for IOCs?** Yes. In project management and systems integration, the Western oil majors still have the skills that the service providers do not aspire to, and which most NOCs are not yet capable of. In finance, the Western oil majors have balance sheets that are bigger than those of the oil service companies and more flexible than those of the NOCs. So they still have a part to play in the riskier aspects of the industry (exploration), the more specialized jobs (liquefying natural gas into tankers) and the tougher locations (the Arctic). It's a shame, however, that these skills are not being put to more use in renewable energy. There are some big offshore wind power projects that would greatly benefit from the oil majors' prime contracting skills and offshore oil experience. But if anything the IOCs are now pulling out of renewable energy. They are either moving to easier projects – Shell, for instance, pulled out of the London Array wind farm in the Thames estuary to invest in onshore wind farms in the US – or scaling down their renewable investment in general, as BP has done. Why? Because, despite their green PR, many oil majors have decided they would rather stick to their highly profitable hydrocarbon business, and leave low-carbon energy largely to utility companies. It is the latter which may one day render the oil majors obsolete.

upside from higher oil prices. They are just service providers. The attraction is mutual. Ashok Belani, Schlumberger's chief technology officer, had the following to say on the subject: "national oil companies can be easier [than IOCs] for us to deal on their home terrain because they obviously know it better than anyone". So a company like Schlumberger, which 20 years ago used to derive 35–40 percent of its revenue from working for the top seven Western oil majors, now draws only 20 percent of its total business from these companies. The vast bulk of oil service company work is now with the NOCs.

Probably only one oil service company, Halliburton of the US, is known outside the industry, and that is largely for non-oil-related reasons. Halliburton was headed by Dick Cheney before he became vice president to George W. Bush, and Halliburton also did a lot of contracting for the US military in Iraq (for which it was accused of overcharging). The company also built the special prison camp at Guantanamo.

Many of the bigger service companies are also US, such as Baker Hughes, which is the third largest behind Halliburton (second) and Schlumberger (the largest), which is a French-American company based in Houston and Paris and stock-listed on both sides of the Atlantic. But some Scottish companies have grown alongside the North Sea oil industry, such as the Aberdeen-based Wood Group, which evolved from a fishing business to an international oil service company, while the other side of the North Sea has produced some major Norwegian oil service companies, such as Petroleum Geo-Services and Aker Solutions.

These Western oil service companies face few rivals from most oil-producing countries, which is why those countries' NOCs have such need of the service companies. Eventually, however, the major Western service-providers will face competition from China as the emerging Eastern superpower, in the form of companies such as the Great Wall Drilling Company.

Utilities

More than just the names on the bills

You're more likely to be in regular contact with the utility company that provides your gas and electricity than you are with your local petrol station, because there are more householders needing heat and light than there are car owners needing fuel.

The usefulness of the expression "utility" is that it covers electricity and gas as well as other basics. It's the most politically sensitive part of the energy industry, because it touches every voter. It's also the one part of the industry that requires some regulation by the public authorities to ensure that energy users are free to choose suppliers, or if they are not free to choose, at least to ensure that they are not ripped off by monopoly energy suppliers.

Even where there is competition at the retail level in utility sales, the electricity or gas still has to be transported from the power station or the gas field via high-voltage wires or high-pressure gas pipes. Because there is not enough space for competing grids, there can only be one set of wires or pipes along a given route, and this constitutes a natural monopoly. And there needs to be a regulator to ensure that this monopoly is not abused.

In contrast to the oil industry – which, because of foreign concessions, was international from the very start – the energy utilities, at least in electricity, had local beginnings. They had often started life as local private enterprises, but after the widespread destruction of World War II in Europe and Japan, they were nationalized by governments keen to plan and accelerate economic reconstruction.

Among industrialized countries, only the US maintained the predominant model of private investor-owned utilities – and even there the federal government went into the energy business in the 1930s Depression by creating the Tennessee Valley Authority, with hydroelectric dams to provide power for poor farmers.

The Fontana Dam: a hydroelectric dam built on the Little Tennessee River. It is operated by the Tennessee Valley Authority – an early US utilities provider – which built the dam in the early 1940s to meet the skyrocketing electricity demands in the Tennessee Valley at the height of World War II.

Moving in the opposite direction

Gradually the utility industry has become a more international business, as barriers to cross-border energy investment and trading have been lowered, first within the European and US markets, and then among some developing countries. This barrier-lowering process – which was known as liberalization or deregulation and which often involved privatization – was launched in the UK, other northern European countries and parts of the US, as something that would boost efficiency and bring consumers choice and lower prices.

The same message was then spread wider by international institutions such as the World Bank to its client governments in developing countries. And many developing countries did liberalize and privatize their energy sectors, thereby opening up investment opportunities for rich countries' utility companies. The welcome that some developing countries were ready to give foreign investment in electricity contrasted with their guarded attitude to foreign investment in their oil.

The backlash

However, many in the NGO community protested that energy privatization in developing countries particularly hit the poor, who no longer had the state looking after their fuel needs. Meanwhile, consumer protection bodies in richer countries complained that the freedom (conferred by liberalization) to choose one's energy supplier was being abused: specifically, some energy companies were using high-pressure sales tactics and misleading information to get consumers to switch supplier. By 2009 the tide seemed to be turning against liberalization in industrialized countries, which were beginning to realize that increasingly important issues, such as energy security and climate change could not just be left to the market to deal with.

In Europe, liberalization meaning the freedom and ability to choose your energy supplier varies across the 27 countries of the European Union. At one extreme there is France, where two companies – Electricité de France and Gaz de France/Suez – each have more than ninety percent of their respective retail power and gas markets, and also own the main transmission systems (despite the fact that consumers there are legally free to choose their supplier, as they are throughout the EU). At the other end of the liberalization spectrum is the UK, which has six electricity and gas companies of roughly equal size (four of them foreign-owned companies, controlled by France's EdF, Germany's Eon and RWE and Spain's Iberdrola, which owns Scottish Power) competing vigorously, by fair and sometimes foul means to take customers off each other.

The European Commission has fought a long battle to introduce minimum conditions for energy competition across the 27 countries. Its particular focus has been to ensure that no electricity or gas company can use any transmission network it may own to block rivals selling energy into its home market. Its particular solution has been to try to "unbundle" networks, turning them into neutral common carriers of energy for all suppliers and all customers.

To this end, the EU has passed three rounds of legislation (between 1996 and 1998, in 2003 and 2009) with each successive law increasing the degree of independence for energy networks. The fact that the EU has had to take three shots at this is indicative of the difficulty of the task. The European Commission's insistence on freedom for cross-border investment and trade in energy, as with anything else in Europe's "single market", has also enabled a few big utilities – EdF and GdF/Suez of France, Germany's Eon and RWE and Italy's Enel – to grow still bigger.

These companies' domination of their home markets is negative, as far as competition is concerned, but their expansion elsewhere and pan-European rivalry with each other is probably a plus for competition overall. Together with Centrica of the UK – the upstream and downstream remnant of the old British Gas monopoly – and some big Spanish renewable energy companies, these companies make up the core of Europe's utilities.

In the US, the fifty individual states may be a part of one country, but in terms of energy market structure you would never know it. They differ from each other far more than the 27 EU states for a variety of reasons, though the most significant is a tradition of states rights and until recently restrictions on financial conglomerates owning utilities.

There are more than three thousand utilities in the US, of which some two thousand are publicly owned – mainly by municipalities in the Midwest and West. More than eight hundred of them are rural electricity cooperatives owned by their consumers. It is the much smaller number of around two hundred investor-owned utilities that make up the core of the industry, accounting for nearly half of generation and two-thirds of sales and profits. They include the likes of American Electric Power, Duke Energy, Exelon and Southern Company.

The conditions under which US utilities operate vary widely. They range from the mid-Atlantic and Midwest states, where largely volumes of power are competitively traded along independently-owned transmission lines serving all comers, to the monopolies (albeit regulated ones), of some southeastern states, where competition from out-of-state is not just unwelcome, it is illegal.

Elsewhere in the US, however, there are still plenty of opportunities for foreign investors. European utilities that are relatively small by European standards (such as the UK's Centrica) loom large in the fragmented US market, while companies that are big by European standards (such as Britain's National Grid, which runs the entire UK transmission system), are positively huge in much of the US. Moreover, European companies

Wind turbines spin in the desert outside Tehachapi, California. Many European companies have found the US to be a much more welcoming prospect than Europe in terms of planning permission.

find it easier to get planning permission to build onshore wind farms in the US than in Europe, where environmental objections are increasingly forcing them to go for offshore wind at greater expense.

Energy and money
Which one makes the world go round?

This chapter takes a look at how the money goes in, out and around the energy sector. It examines the relative subsidies to fossil fuels and renewable energy, and looks at the controversies over whether financial speculators force up the price of oil or whether the energy companies rip off their customers.

The money that goes in

Tourism is often claimed, at least by those inside it, to be the world's biggest industry. But sightseeing, almost by definition, depends on transport, which is but a sub-sector of energy. The overall amounts of money now going into energy are now huge. The International Energy Agency reckons that cumulative investment in all kinds of energy to the tune of $26 trillion (or $26,000,000,000,000) will be needed in the period from 2007 to 2030. That estimate was based on the value of 2007 dollars, not allowing for any future depreciation of the greenback. It is also based purely on current projected needs, not allowing for any new crash climate-change programme that may well be required over the next two decades.

Just over half (nearly $14 trillion) of this huge amount would go to the power sector – new generation and distribution systems for electricity. These are getting much more expensive, especially if countries want to opt for low-carbon nuclear reactors – the cost of the new generation of French reactor has risen to €5bn – or wish to avoid unsightly onshore windmills by putting turbines and grid connections into the sea. Indeed, virtually all the cost of many renewable energy systems will come in the form of big upfront costs and with subsequent costs for maintenance (although with zero costs for fuel, in the form of wind or sun). Moreover, the uses of electricity are also going to expand. Providing that we can find new ways of generating it carbon-free, this is a good thing. Powering cars, for instance, is one potentially enormous new use for electricity.

But the part of the energy industry with the highest financial profile is undoubtedly the big oil companies. Their offshore platforms are the most spectacular engineering part of the energy industry. Until the development of big wind turbines, their service stations made the biggest visual impact on our landscape of any commercial sector. And the tax on their petrol sales is a staple of government budgets.

Big spenders

Most of the companies' capital expenditure (or "capex", as it's known) is upstream – meaning exploring for and producing oil and gas out at sea, in the frozen tundra or in the desert – because that is where their biggest profits are. This expenditure upstream more than tripled, according to the IEA, from an annual rate of $120bn in 2000 to $390bn in 2007. Nearly two-thirds of this was spent by the international oil companies (as distinct from the national companies, such as Saudi Aramco, that are either state-owned or operate just within their national boundaries or both) and half of that was spent by just five companies. These are the so-called supermajors – ExxonMobil, Royal Dutch Shell, BP, Chevron and Total. They acquired this grandiose title in recognition of how they grew during the late 1990s when the oil price hit a trough of $10 and set off a wave of mergers between companies (which were already pretty big). Exxon merged with Mobil, BP with Amoco and Arco, Chevron with Texaco, and Total with Petrofina and Elf.

Only Shell was big enough already to stay in the top five without a major acquisition. After a scandal in 2004 about overstating its reserves, however, the half-marriage dating from 1907 of Royal Dutch Petroleum and Shell Transport and Trading of the UK was finally consummated and the two firms were rationalized into a single integrated company called Royal Dutch Shell.

In 2008 the combined capex of the Big Five was $106bn. Where does this money come from? Mainly from the companies' own balance sheets. The oil companies take in huge amounts of money – the Big Five produced a combined net profit of $136.49bn in 2008 – but they also pay out huge amounts on new projects. When the big oil companies have to borrow on the capital markets to top up their own funds, they are able to borrow general-purpose funds without having to tell lenders exactly what the borrowing will be spent on.

Small oil companies, without the credit rating or track record of the oil majors, usually have to get specific finance for each project, and to pledge future production from the oil or gas field in question in order to under-write their borrowing.

Risk-sharing

For all the industry's recent technical advances, finding commercially extractable amounts of oil and gas is still a hit-and-miss business. So the oil companies have taken to splitting risks and expense by partnering with each other on upstream oil and gas projects, despite the fact that any such collaboration in *downstream* petrol and oil sales and prices remains absolutely illegal under anti-trust law.

To take a striking example, the Kashagan offshore oil field in the Kazakh zone of the north Caspian is being jointly developed by Eni of Italy, the Anglo-Dutch company Shell, ExxonMobil and ConocoPhillips of the US, France's Total and Inpex of Japan as well as the local Kazakh oil company. The idea of, say, US, European and Japanese car companies sharing a joint car factory in such a way would be unthinkable. Car companies would worry about their partners spying on their proprietary design and marketing, which of course is far greater in a manufactured product than a simple commodity.

It is true the oil companies have developed some proprietary tech-nologies, but this is supplemented by techniques and services supplied to the industry on a common basis by the oil services companies like Halliburton and Schlumberger – and there is nothing very special about the end-product of oil and gas molecules.

Renewable energy finance

Renewable energy is growing fast from small beginnings, because what is special about the electrons produced by wind and solar power, or bio-gas molecules, is that governments are allocating them increasing amount of subsidy and support in the fight against climate change. The green stimulus portion of governments' 2009–10 economic recovery plans (see p.20–21) boosted the amount of money going into clean energy research,

An aerial view of an artificial island on the Kashagan offshore oil field in the Caspian sea, western Kazakhstan. The whole operation is being run by an international consortium of different oil companies.

development and deployment by up to $80bn (over several years) in the case of the US.

With the exception of the nuclear field, research and development spending in energy, by both the public and private sector, compares poorly with areas such as defence and pharmaceuticals. In terms of ongoing production and consumption subsidies, the overall level of government subsidy is hard to calculate, because it comes not only in cash payments, loans, loan guarantees and tax credits but also in regulation with a financial impact. For instance, renewable energy producers benefit financially if customers are required to buy a minimum quota of green electricity, because this quota raises demand (and perhaps prices) for their green electricity and lowers the risk (and perhaps the cost of capital) of investing in it. But the cost of this regulatory support is borne by consumers rather than taxpayers.

The really huge energy subsidies in today's world are still for fossil fuels. These are estimated as being as high as $300bn a year, partly in tax breaks in countries like the US for oil and gas production, but mainly in subsidizing the cost of oil products for consumers in developing (often oil-producing) countries. The 2006 Stern Review estimated that global direct government (taxpayer-funded) support for low-carbon energy at $26bn a year – but $16bn of this went to nuclear and $10bn to renewables. A study of US energy subsidies between 2002 and 2008 by the Environmental Law Institute put the amount going to fossil fuels at $72bn, compared to $29bn to renewable energy (of which $16.8bn went to ethanol alone).

As for investment by the renewable energy industry itself, many renewable energy companies are naturally still in the incubation phase, growing in some university or government lab and hoping to attract the attention of some venture capitalist or private equity investor. But many others have taken wing and are flying solo. According to *New Energy Finance*, there was a trend, before the 2008–9 retrenchment, by European utilities to float their renewable energy operations as separate, free-standing ventures rather than as part of a larger parent. This, commented *NEF*, showed "a belief that capital markets will rate 'new' energy higher than 'old' energy, because of its higher growth rate".

The high point of this trend was the success of Iberdrola, a Spanish utility, managing to profitably sell off 20 percent of its Iberenova renewable energy subsidiary for €4.1bn at the end of 2007. The 2008–9 recession has interrupted investment in renewables. But resumption of growth, when it comes, will be from a fairly high level. According to *NEF*, capital expenditure in renewables – internal and external sources of finance – totalled $136bn in 2008, of which nearly $100bn was in new projects.

The future pace of private renewable investing will depend very much on the degree to which governments underwrite it by regulation or money. In September 2009, in the run-up towards the Copenhagen climate summit, a group of 181 institutional investors, representing $13 trillion in assets under their management and including such big names as HSBC, Hermes and Swiss Re, called on governments to use both instruments. They argued for stringent carbon emission cuts of between 50 and 85 percent by 2050 in order to level the global playing field and to provide some kind of investment certainty far into the future. In addition, they suggested government guarantees for their investments.

The money that goes around

Energy trading has become a very big business. And to many who believe that speculators force up energy prices and/or encourage energy price volatility, it has also become a very suspect business. For the price of oil is of obvious importance to the world economy as well as being something that is influenced by politics – and which in turn influences politics.

Energy sources – chiefly oil but also gas, coal and electricity – are relatively recent additions to the list of commodities, predominantly agricultural, that have long been traded. It is no surprise that energy has come to be traded as a way of matching supply and demand on the spot market (for immediate delivery), and of reducing the risk of future price changes on the futures market (for future delivery). Energy is fairly easy to standardize, making it easier to trade on an exchange, and (except for electricity), it is easy to store and therefore hold for future delivery. But the reasons why it should have become such a major feature of commodities trading lie in other developments in the history of energy.

Oil

When internal oil exchanges among the Seven Sisters disintegrated under the impact of OPEC nationalizations, they were replaced by an open trading system. Up until the 1970s, most of the world's oil had flowed through the Western oil majors, which were vertically integrated operations: from upstream concessions, via mid-stream refineries, to downstream petrol stations. This vertical integration was blown apart when OPEC governments nationalized the Western companies' upstream concessions, leaving them short of crude for their refineries and petrol stations. Eventually they found new sources of crude outside OPEC, from new fields in the North Sea, Gulf of Mexico and Russia. But now the big difference was that, instead of the connections between the upstream and downstream being in secretive telexes between the Seven Sisters, they took place in trading pits on the New York and London oil futures exchanges. Oil had always been traded in some form ever since it was discovered 150 years ago, so this was nothing new. But what was once private became very public.

Gas and electricity

From the 1990s, liberalization has had a similar impact on the traditional vertical integration of utilities in many parts of Europe and America. Liberalization – encouraging everyone to compete against everyone else – destroyed the old certainty of vertical integration. Up until then, companies could always be sure that no one would poach their retail customers, and therefore found it relatively easy to match supply to these customers' demands. But liberalization tended to throw supply and demand out of kilter – and so the need to re-balance supply and demand with trade grew.

The new ways of trading

As the old system of administrative setting of prices first by the Seven Sisters, then by OPEC broke down definitively in the mid-1980s, trading was mainly on a spot basis – sellers had to deliver and buyers had to take delivery immediately. This naturally limited trading to those with oil and those with a use for oil or storage for oil. But futures trading quite quickly developed, partly for the usual reason of reducing price risk and partly because the Western oil majors needed a price benchmark to show to their tax authorities (which were now able to take a more active interest in the newly opened-up internal workings of the oil companies) in order to prove their balance-sheets were legitimate.

In contrast to spot transactions, trading oil futures is theoretically open to an infinite number of players – far beyond the necessarily limited number of people with a need or capability to buy, sell or store physical oil. In particular, futures trading brings in the speculators – people whose only interest in physical oil is what they put in their car or lawn mower but with a yen for a bit of risk-taking. That is what makes speculators, despite their popularly disreputable image, the vital complement to hedgers in futures markets.

Hedgers and speculators

Hedgers are people, usually producers and users of oil, who want to avoid the risk of the price of that oil changing in the future. A hedger could be a small oil company which, to underwrite a bank loan it has taken on to develop a new project, has had to guarantee to the bank that it will be able to sell future oil from the project at, say, $70 a barrel four years hence. In "selling its oil forward" at $70 in four years time, the oil company is willing to forego the possibility of the market price for oil going *higher* than $70 in four years time, just in order to be sure its own oil will fetch no less than $70.

Equally, a hedger might be a *consumer*, rather than a producer of oil. A chemical company, say, might wish to ensure that it will not have to pay more than $70 a barrel for its crude oil feedstock in four years' time. Again, the chemical company is willing to pay a possible price to gain certainty – the price being the possibility that the market price for crude might be below $70 in four years' time and that if the chemical company had not hedged its oil purchase it could have got its feedstock cheaper. In sum, hedgers are willing to give up the opportunity to benefit from favourable price changes in order to protect themselves from unfavourable price changes.

Speculators are the precise opposite: they hope to profit from the very price changes that hedgers seek to avoid. That is why they will buy what hedgers want to sell, or sell what hedgers want to buy. So a speculator might buy a future contract for oil at $70 from the hypothetical company above in the hope or expectation that they could sell the oil on for more than $70 within the four years. They could equally buy a futures contract, or the right to sell that chemical company oil at $70 in the hope or expectation that the actual cost of supplying the chemical company will be less than $70 four years hence.

So the speculator and hedger are a useful, indeed necessary, match for each other. They don't have to be an *exact* match; it is just useful to have enough speculators to make deals with hedgers whenever the latter want to cover their risks. The main information we have on the relative weight of these two groups comes from surveys by the Commodity Futures Trading Commission, which regulates the US futures exchanges.

According to the CFTC, hedgers – whom it defines as people with a physical interest in oil – are on average roughly two-thirds of all market players, with "non-commercial participants", or speculators, the remaining third. When there are enough participants so that it is easy to quickly find a price to make a deal with, the market is said to be "liquid"; when there aren't enough the market is said to be "illiquid".

Driving up the price?

One of the main oil-market trends of recent years has been the increase in spread betters, who speculate on differences between futures prices in different months, rather than on the overall future level of crude oil prices. But there is also no doubt that so-called "momentum speculators" have entered the market to drive the price up or down faster than any reasonable examination of the fundamentals of oil supply and demand would warrant.

Indeed, such bandwagon speculators are not interested in boring details like the fundamentals. They just follow momentum strategies that dictate buying whatever is going up in price or selling whatever is going down. This increases volatility – the way in which the oil-price jumps around. Oil is not the most volatile of energy products: the price of electricity can vary betwen three hundred and four hundred percent during a day's trading because it's not storable, yet it's something that people usually need *immediately*, whatever the cost. But oil is nevertheless by far the most important of traded energy products. So the world really felt it when crude oil rose briefly to a record $147.27 a barrel on 11 July 2008 on the New York Mercantile Exchange. On the way to that record high, there had been the largest daily increase ever in the oil price, of $10.75 on 6 June of that year. That daily increase was higher than the total price of a whole barrel ten years earlier. Likewise, on the way down there was a record drop for one day, of $14.31 on 23 September.

The run-up in the oil price to mid-2008 and then the seventy percent plummet in the price by the end of 2008 has prompted calls, particularly in the US and particularly in Congress, for tighter curbs for speculative oil trading and greater transparency in the futures markets. Some changes are likely, and may be mirrored in Europe, even though a US inter-agency task force concluded in June 2008, as the oil price was reaching its peak, that "current oil prices are being driven by fundamental supply and demand factors".

Oil benchmarks

World oil-trading revolves around three benchmark products chosen for high quality and convenient location. They are "West Texas Intermediate", or WTI, traded primarily on the Nymex in New York; "Brent" (originally from the North Sea's Brent field, but now a mix) traded primarily in London, and "Dubai" traded in – you guessed it – Dubai in the Gulf. The money spent on trading futures in these products is enormous. Take the Nymex (New York Mercantile Exchange):

▶ The daily average volume of crude oil trades for 2008 was 532,309.

▶ Each crude oil contract is for 1000 barrels.

▶ The average daily price during 2008 was $99.73.

So the amount spent trading crude oil on an average day on the Nymex in 2008 was $53.087 billion (532,309 x 1,000 x $99.73). Not exactly peanuts.

Another way of illustrating the magnitude of oil trading is to look at the balance sheet of BP. In contrast to ExxonMobil (which just tends to trade its own oil), BP sees trading other people's oil as well as its own as a general opportunity to make money. BP's sales turnover for 2008 totalled $361bn. Of this, $44bn came from upstream production operations (usually the most profitable part for the big oil majors), and all the rest – $317bn – was from downstream operations such as refining, marketing,

The small Oklahoman town of Cushing is the self-proclaimed pipeline crossroads of the world.

chemicals, of which trading was an undefined part, but almost certainly the largest.

This turnover figure is swollen by inclusion of cargos of oil sold over and over again. Indeed, it is distorted to the point of meaninglessness, so that professional oil-industry analysts pay little heed to the oil majors' sales figures. Nonetheless, it is striking to compare this with Microsoft a huge company but in this context a "normal" one with in 2008, a sales turnover of *only* $60bn.

Yet even more extraordinary to the layman is that less than one percent of these trades – averaging more than half a million a day in 2008 – actually results in the physical delivery of oil. The reason is that futures contracts, such as Nymex's WTI contract, are good for gauging price, but bad for any practical allocation of oil.

All features of a futures contract are standardized, with notional delivery for WTI set at Cushing, Oklahoma – except for price. Because everything but price is standardized, these contracts are an ideal medium for getting a good view of price through trading. But in terms of allocating oil, the delivery feature of these contracts is totally impractical: very few people want to deliver, or take delivery of, oil in a small town in Oklahoma.

But world oil-trading not only needs to provide flexibility of location. It also has to cater for dozens and dozens of different grades of oil, which is far from a homogenous product. So what has developed is a vast web of bilateral contracts in what is known as over-the-counter (OTC) trading. These OTC contracts take WTI, Brent of Dubai (which are really just idealized abstractions) as price benchmarks, and price the oil in question at a discount to reflect quality, transport and destination. It is really through OTC trading that most of the world's oil is allocated and distributed. However, buyers and sellers of OTC contracts will usually use the three main futures markets to hedge their risks and exposures on OTC contracts. This explains how only one percent of trades on the futures exchanges result in actual delivery, and yet two-thirds of those trading on these futures exchanges can be classed as having a physical interest in oil.

Oil dwarfs trading in other energy commodities. But these are not negligible. For instance, about a third of Germany's large electricity consumption is traded on Leipzig's European Energy Exchange, where over €85bn changed hands in 2008. The UK trades heavily in energy – about the same as Germany in electricity, but Britain's daily gas trade of more than £500m is around seven to eight times greater than in all of the rest of Europe, where gas is typically sold on long-term contracts indexed to the oil price. In addition, there is now the new "commodity" of carbon which is traded under the European Union's Emission Trading Scheme. In 2008 global carbon trading transactions totalled $120bn, *New Energy Finance* has estimated.

The European Energy Exchange is based in Liepzig, Germany, and operates market platforms for trading in power, natural gas, greenhouse-gas emission allowances and coal.

The money that comes out

Energy companies can generate huge profits. For 2008 the biggest five oil supermajors reported a collective net profit of $136bn, and no wonder. For some of the oil they could sell in that year for over $140 a barrel came from fields that were planned and drilled when the oil price was $20 or less a barrel. This is the sort of profit level that tempts politicians to levy windfall profits taxes on the oil industry. However, it's worth remembering that the capital requirements of the oil industry are also huge – the same Big Five shelled out more than $100bn in capital expenditure in 2008 – and the oil price is fantastically fickle.

No wonder, then, that governments around the world regard the oil industry as a cash cow to be milked. The UK, like most other oil-producing countries, taxes oil corporate profits at higher rates (50–75 percent, depending on the oil field) than the rest of the corporate sector – as compensation for the fact that oil and gas are a finite resource, which both increases their value over time and makes them irreplaceable. Over the last forty years the UK treasury has gained £271bn (in 2008 money) in corporate oil tax.

Excise and value-added tax on petrol also bring in a huge amount of money, especially in Europe and Japan less so in North America where gasoline tax is much lower. Tax accounts for more than half the pump price of most European petrol, a fact that once led a Kuwaiti oil minister to remark that he would be prepared to give European governments his oil free, if they would just split their petrol-tax receipts with him.

Most investors in the oil industry like a bit of a flutter, either on the oil price taking off or the company in question striking it big. Only real gamblers, however, should invest in the smaller oil and gas companies, often listed on London's Aim market, which frequently have just a couple of oil wells or gasfields.

Energy investors of a more nervous disposition should consider electricity or gas utilities, particularly those with a large portion of regulated assets such as networks, electricity grids and gas pipelines. Regulation, usually designed to ensure that network monopolies are not abused, typically provides companies with a guaranteed return on their assets and capital expenditure.

In the end, it is the consumer who has to shoulder the higher energy prices and taxes (though some of the revenue is spent on public services, of course). Consumers often feel they are being over-charged by energy companies. Sometimes they are genuinely being ripped-off, by companies that are quick to pass on to the consumer rises in wholesale energy prices, but slow to let consumers benefit from falls in wholesale prices and awkward in letting consumers switch supplier.

Yet part of consumer frustration with energy arises from exaggerated expectations encouraged by politicians of the price benefits of liberalization and from a reluctance (also encouraged by politicians) to accept the need for energy prices to rise to reflect carbon's real social cost to our environment.

PART 4
ENERGY
AND
EMERGENCY

How close are we to
the edge?

Security of supply

The serious business of keeping a nation switched on

Absolute energy security, like immortality for humans, is unachievable. Too much can go wrong in the way of conflict, natural disaster or technical failure for total reliability to be possible.

But governments know they cannot afford to adopt a fatalistic, passive attitude to the possibility of their electorates freezing in the dark. And risks of interruptions in oil and gas supplies are increasing because of:

▶ a continued rise in oil demand for transport among virtually all developing countries, while climate-change concerns drive a demand for relatively clean natural gas.

▶ insufficient investment in oil exploration and production, because national oil companies lack the incentive to increase capacity, but refuse to give Western oil majors – which have that incentive – access to their reserves.

▶ an ever-greater concentration of oil reserves and therefore of production in the hands of fewer countries. This means more energy passing through various chokepoints. In 2006, forty percent of oil transited through the four conduits of the straits of Hormuz, Malacca, Bab-el-Mandeb and the Suez canal. By 2030 this ratio will be sixty percent, the IEA estimates.

Any disruption in oil flows has an immediate ripple effect in price terms throughout the world oil market. So a serious shortfall of supply to one country, or in one country, does not necessarily result in a physical shortage of oil for everyone else. But because oil is one big, interconnected market, such a shortfall results in a higher oil price for all countries. By contrast, disruptions in gas supply rarely have any global effect. For instance, the total cut-off in the flow of Russian gas across Ukraine to Europe in January 2009 had barely any effect on North American or Asian gas markets. The world gas market is a series of regional markets, fed mostly by pipelines and only thinly interconnected by the relatively few tankers able to ship gas around the world. The gas market, built largely on fixed infrastructure and long-term contracts between buyer and seller, is segmented. Crises in gas supply tend to be local or regional, requiring local or regional solutions.

The International Energy Agency: be prepared

Because the disruption of oil supplies has a worldwide price impact, there is a clear advantage in international cooperation to minimize this impact. This was the rationale for the 1974 decision by major oil-consuming countries to set up the International Energy Agency. It was partly a reaction to the move by the Arab members of OPEC (partly related to the Arab-Israeli war of October 1973) to cut their oil production by 25 percent and to start fixing oil prices with a 400 percent hike above previous levels.

Over the years, the IEA has developed into a forum of discussion and study of all aspects of energy, including energy market liberalization, technology, efficiency, renewables and climate change. But at its heart is its Emergency Response Mechanism – the twin commitment of IEA countries to hold oil stocks equivalent to at least ninety days of net oil imports and, in the event of a major supply disruption, to release some of the stocks on to the market, as well as to take other measures to restrain demand and, if practical, switch fuels and increase domestic production.

The IEA has only acted twice to put these emergency procedures into effect. The first occasion was in January 1991, when IEA members put some 2.5m barrels a day on the market to prevent the start of the Gulf War creating a spike in the oil price. The second was in 2005, in the aftermath of Hurricane Katrina. In addition to flooding New Orleans, Katrina not only stopped some production in the Gulf of Mexico for weeks, but – even more importantly – knocked out most of the refineries along the Gulf coast. So IEA members agreed to draw on their stocks to put some 2m b/d on the market – not only of crude oil but also refined products.

The US Strategic Petroleum Reserve

As the world's biggest oil importer – roughly 12m barrels a day – the US holds the biggest publicly-owned emergency stock of any IEA member. It holds more then seven hundred million barrels in four underground salt caverns along the Gulf coast (two in Louisiana and two in Texas) and has voted legislation to bring this up to one billion barrels. In addition, US oil companies hold an almost equally large amount as part of their commercial stocks.

A technician at the Strategic Petroleum Reserve inspects a crude-oil transfer pipe. These are in some ways America's most important pipelines: the fall-back in a real crisis.

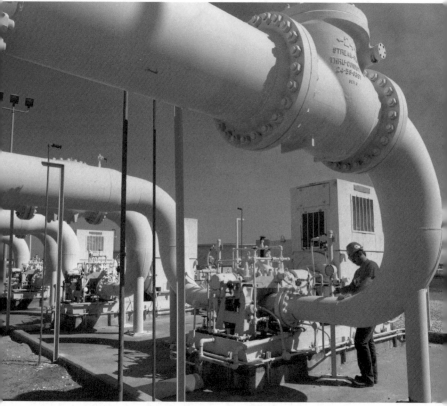

Major oil supply disruptions

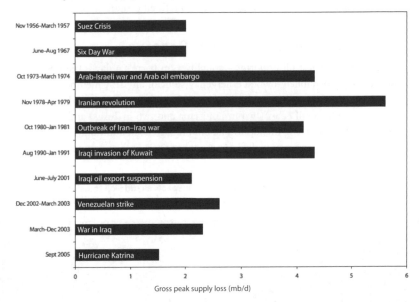

Source: International Energy Agency

Apart from taking part in IEA-coordinated action, the US has only drawn on the Strategic Petroleum Reserve on one other occasion. In 2000 the Clinton administration authorised the temporary release of some oil, on the condition that the same amounts were subsequently returned to the SPR, in order to "cool" the rise in the oil price.

The action had the desired effect of dampening the oil price. But Bill Clinton's action was criticized by his successor as playing politics with an SPR that was only supposed to be used in case of a physical shortfall in the oil supply. In the wake of the terrorist attacks of 11 September 2001, and in the context of his war on terror, President George W. Bush started to fill the SPR further, up to a maximum of one billion barrels.

The European Union and gas storage

Of the 27 EU member states, 19 are members of the IEA and some of the rest will join in due course. (In order to join the IEA, countries first have to belong to its parent body, the Organisation for Economic Cooperation and Development, the Paris-based club of market-oriented economies.) The EU has its own legislation on oil stocks (which is important for the eight EU members that do not belong to the IEA), but this legislation is very closely modelled on the IEA oil-stock rules.

However, Europeans have focused their energy-security concerns less on oil and more on gas – and on Russian gas in particular. The EU is surrounded by a variety of suppliers within pipeline distance (such as Norway, Azerbaijan, Libya and Algeria), and by others (such as Qatar and Nigeria), which can ship their gas to Europe by LNG tanker. However, Russia has a higher share of world gas reserves (25 percent) than Saudi Arabia has of world oil reserves (21 percent), so it is likely to be Europe's mainstay gas supplier for the long term. Provided, of course, that the rocky relations between Russia and Ukraine – still the conduit for eighty percent of Russia's Europe-bound gas exports – do not bring Russian gas flows to a halt, as happened briefly in 2006 and more seriously in 2009.

The IEA sets no rules on gas storage for its members, partly because gas is, for reasons mentioned above, not an issue of common concern to its membership, and partly because gas is harder and costlier to store than oil – only some countries have the right geology to do so on a large scale. In fact, the IEA has advised against any set requirement for strategic gas stocks. ("Strategic" stocks refers to stocks designed to fill the gap in a

A Ukrainian demonstrator gestures during a protest against Russian plans to increase the price for gas in front of the Russian embassy in Kiev in late 2005.

Several dozens of Ukraine's nationalists staged a protest as Kiev and Moscow launched a war of words over natural gas supplies for 2006, which ultimately culminated in Russia briefly turning off the Ukraine's gas supply (as it did again again in 2009).

Brokers swarm and signal in a frenzy of buying and selling at the Tokyo stock exchange in December 1973, following reports of an Arab decision to ease its oil-supply cutbacks against Japan – cutbacks which had seriously affected its economy.

major supply cut-off, rather than merely to smooth out the usual seasonal swings in commercial demand.)

Nonetheless, the 2009 cut-off of Russian gas has galvanized the EU into some action to improve the interconnections between its member states' gas networks and commercial storage, as part of a series of measures to improve Europe's resilience to interruptions to outside supplies.

Japan: the ultimate importer

Japan has built itself into a major industrial power without ever having any real oil of its own. It is the third-largest oil importer, behind the US and China. Yet it has made great efforts to contain its oil demand, which is essentially unchanged from 35 years ago.

This is because it is, and has always been, worried about the reliability of oil supplies. One factor here is that ninety percent of Japan's oil comes from the Middle East, and of that fully one third is from Saudi Arabia. Japan was an enthusiastic founder member of the IEA, precisely in order to gain some international solidarity on energy security.

Big new consumers: China and India

With the emergence of China and India, the IEA is no longer the predominant oil consumers' club. Virtually all the growth in oil demand these days is coming from developing countries, especially China and India. By the IEA's own reckoning, non-IEA consumption for oil will match that of all the OECD countries by 2015 and be substantially higher by 2030.

The IEA realizes that energy security is indivisible, in the sense that panic-buying by countries the size of China and India could have big impacts on the oil market and set off a scramble for oil resources. So the IEA has for some years had an "outreach" programme towards Beijing and Delhi, especially to persuade them to start building up their own strategic oil reserves. Whether or not at the IEA's persuasion, China and India have both started to create oil stockpiles, which might reach five hundred million barrels for China and one hundred million barrels for India.

But it seems unlikely that China and India will ever join the IEA outright, and not only because of the obstacle that OECD membership is a precondition for joining the energy agency. For neither China nor India may be ready to take on the collective obligation to put oil stocks on the market in an emergency. Both countries seem to put more faith in their own energy security efforts – whether to build their own stocks purely for their own emergency use, or to acquire equity stakes in foreign oil fields.

Fuelling the Chinese dragon

China became a net oil importer in 1983. By 2006 it was importing 3.7m b/d or about fifty percent of its total consumption. In the absence of any major change in policy, China will be importing over 13m b/d in 2030, or 80 percent of its consumption. The vast bulk (eighty percent) of China's foreign oil purchases comes from the Middle East and Africa, and therefore has to transit the narrow Straits of Malacca (see p.205).

Caution: oil can damage your state

Oil resources can be a blessing to states that use them wisely. The model of a prudent petro-state is Norway, which has not gone on a spending splurge, but put aside surplus oil revenue for future generations of Norwegians (for the rainy day when the oil and gas eventually run out). But then Norway had fully developed as a state before it ever found oil. For all those countries where oil exploitation and state-building still go hand in hand, oil wealth ought to carry an economic and political health-warning.

The first drawback is economic. The problem is that a state's export of a valuable natural resource like oil or gas can, by raising the real exchange rate, make other goods less competitive, and so over time de-industrialize the country in question. This phenomenon was dubbed "the Dutch disease" after large gas finds in the Netherlands seemed to lead to a manufacturing decline. Such industrial decline would not matter, provided the state could go on exporting the natural resource forever. But while the Dutch can probably go on exporting tulips forever, their gas will run out.

While this problem can occur in sophisticated economies and long-established political systems, like the Netherlands, the second drawback of oil wealth applies specifically to developing countries. This is the problem that resource riches can warp or stunt political institutions. The argument goes that, in such petro-states, the ease of extracting revenues from natural resources means that governments do not need to do the hard work of creating the public services that would justify taxing their citizens, because they don't need to tax them. The logic is that petro-wealth breaks the link between taxation and state-building on which, for instance, state formation was based in Europe (whose monarchs had to go to parliaments to raise taxes in order to fight their frequent wars).

Certainly oil wealth lessens the pressure on conservative rulers of the Gulf petro-states to give their citizens political representation. If "no taxation without representation" was the slogan of the American Revolution, the reverse can also be true – "no need for representation if no taxation". Even petro-states with a more developed democracy, such as Nigeria, Venezuela, Iran and Indonesia, tax their non-oil sectors more lightly than countries without oil, because they feel they can afford to.

Other features of oil can mean that a state with oil may be no better off than one without. In developing countries, the oil sector is typically an economic

China is pursuing a twin-track policy to improve its energy security. In addition to building a domestic emergency petroleum reserve, Beijing has directed its three national oil companies to undertake what they call a "going out" policy – to take equity stakes in oil fields around the world in order to diversify and secure sources of supply. Like India, China appears fearful that the international market might not make enough oil available island. It employs lots of capital (often foreign), relatively few local workers and is often fairly isolated from the rest of the economy.

But can oil wealth actually make countries worse off? It can, argues Terry Lyn Karl, a US academic whose book *The Paradox of Plenty* is one of the best-known works on this issue. For a start, oil wealth can be more corrupting than other forms of wealth because it arrives more directly into government coffers.

Oil revenue or royalties flow to the government in states everywhere – even in the US, which is the only country to allow private ownership of the subsoil (including oil resources). But in states that have never developed any tradition of fiscal accountability towards their citizens – typically in low-tax petro-states – government officials may tend to help themselves to state oil-income. Oil wealth also appears to provide funding or incentives for civil wars and military or political coups.

Oil-price bubbles can leave a country worse off once they burst. The oil booms of the 1970s certainly left many oil producers worse off. Their spending splurges and ballooning subsidies caused inflation, surging foreign debt and much higher debt-to-export ratios than non-oil producing countries. More importantly according to Karl dependence on petroleum – the one fate exporters wanted to escape – increased markedly after the boom.

"Only one boom in history – that resulting from the discovery of gold and silver in the Americas – rivals the 1973 and 1980 oil bonanza," she writes, and what happened to Spain in the 16th century is not a good precedent. Imports of New World bullion drove up the value of Spain's currency, encouraged inflation and imports, and depressed agriculture so that by 1570, Spain was having to use coin from the Americas to pay for imported food. Spain's version of the paradox of plenty is summed up in the words of one of its proto-economists of the time: "if Spain has no gold or silver coin, it is because she has some; and what makes her poor is her wealth".

Luckily, today's oil producers generally appear less profligate than Spaniards of the 1570s – or indeed their counterparts of the 1970s. Taking a leaf out of Norway's book, oil exporters such as Russia and Azerbaijan have created oil stabilization funds.

to it a crisis, as other countries might intervene to divert the physical flow of oil. This perception was reinforced by the US's political resistance to the Chinese National Offshore Oil Corporation (CNOOC) buying the American company Unocal in 2005.

To this end, Chinese companies have acquired stakes in Kazakhstan, Russia, Venezuela, Canada, Sudan and West Africa. In particular, after

Chinese building workers at the construction site of the Al Mogran Project in Khartoum, Sudan, in January 2007. The site is set to be the headquarters of both the Greater Nile Petroleum Operating Company and Petrodar – two major oil companies that are joint ventures between China, Malaysia, India and Sudan.

being snubbed over Unocal, Chinese companies have moved into Africa. Nine of China's ten top trading partners in Africa in 2008 were oil-producing countries. (The odd one out is South Africa, which has other minerals to offer the Chinese.)

Energy, war and US policy

The American mix

Oil has been the lifeblood of war – armies, navies and air forces run on it – ever since the major powers' navies converted to oil just before World War I. Oil fields have therefore been a prime target in war. During World War II, getting hold of the oilfields of Romania and southern Russia was a vital goal for the Germans, and loss of them after 1944 helped seal Germany's fate. When US and UK troops invaded Iraq in 2003, one of their priorities was to seize Iraq's oil fields.

When US troops took Baghdad, it was noticeable that they secured the oil ministry, while leaving the other ministries to be looted by Iraqis. (Leading Donald Rumsfeld, the US defense secretary, to make his famously insouciant remark about the looting: "stuff happens".)

But just because oil fields are a prime target in war does not mean that they are necessarily a prime *cause* of war. Oil certainly can, and has been, the trigger for war. It was the reason why Bolivia and Paraguay went to war in the 1930s over the Chaco region, which was, wrongly, thought to hold oil. Many historians consider oil to have been a motive behind Japan's attack on the US in 1941. The Japanese attack on Pearl Harbor appeared to be in response to the decision of America, then the world's major source of oil, to ban the sale of oil to Japan, in order to penalize Japan for its incursions into China. The Japanese had calculated that the only option left to them by the US decision was to invade the Dutch-controlled Indonesia to get oil, that this would make war with the US inevitable, and that therefore they might as well land the first blow, at Pearl Harbor.

The US and the Middle East

In 2003 many people were convinced that the US was invading Iraq primarily for its oil, and many protestors against the invasion carried "No Blood for Oil" placards. Donald Rumsfeld said at the time: "this is not about oil, and anyone who thinks that is badly misunderstanding the situation". It was a denial that, given his position at the time, did not carry a great deal of weight for many people.

However, the evidence suggests that the 2003 invasion was driven somewhat more by neo-conservative ideologues wanting to make the Arab world "safe" for democracy and, incidently, for Israel than directly by a desire for oil. If anything, oil was more of a driving force in the first Gulf War, both for Saddam Hussein's occupation of Kuwait in 1990 and for his expulsion from Kuwait in 1991 by the US-led coalition.

But, energy still looms very large in US foreign policy in the Middle East. This is the subject of much of this chapter. For most of the recent history of the link between energy and war turns on the relationship between the US – as the world's largest user and importer of oil – and the Gulf region as the world's most concentrated repository of oil and gas. The Gulf produces nearly forty percent of the world's oil and almost a fifth of its gas, and is home to a still larger proportion of world reserves in both oil and gas.

For the US, oil has always been a public-private partnership. Unlike most countries, the US has never had a national state-owned oil company, but its public authorities have always been closely involved in regulating oil, and to some extent in promoting its use. For most of the twentieth century, regulation of production (to prevent over-production) was conducted at the state level – by the two main producing states of Texas and Louisiana individually, and by the two states together in the Interstate Oil Compact Commission. Protection from imports (to prevent US producers being undercut by cheaper imports) was provided at the federal level – by an oil-import tariff imposed in 1932 and by oil-import quotas that lasted from 1959 until they were removed in 1973.

America entered World War II as the world's leading oil producer, believing it possessed about 20bn barrels of oil, or nearly half the world's known oil reserves. But this belief caused as much anxiety as assurance, because it meant that if and when the US went through its own reserves there were few other suppliers it could turn to. The galloping fuel consumption of the Allies during World War II increased this anxiety. So the US began to eye the Middle East as a source of oil.

Alexandria, Egypt in February 1945: President Roosevelt (seated, right) confers with King Ibn Saud of Saudi Arabia (seated, left) aboard an American warship at Great Bitter Lake. The historic meeting took place while the President was en route back to the US from the Yalta Conference.

Before the war, US companies had acquired oil stakes, alongside British and French rivals, in Iraq and Kuwait. But the most important was the sixty-year concession in Saudi Arabia that the Standard Oil Corporation of California (Socal, which is today part of Chevron) bought in 1933.

Socal discovered oil in Saudi Arabia in 1938 and a year later was exporting it. It is a measure of the US government's increasing oil angst that in 1943 a plan was discussed in Washington to create a state-owned company, dubbed the Petroleum Reserves Corporation, to buy part of Socal's oil and set it aside for military use.

This never came to anything. But the growing relationship between the US and Saudi Arabia was cemented at the highest political level in 1945, when President Franklin Delano Roosevelt flew to Egypt and met King Abd al-Aziz Ibn Saud on a US destroyer in the southern mouth of the Suez canal. This was the basis of America's ongoing oil-for-security relationship with Saudi Arabia.

To carry out its security side of the bargain, the US were given the right to set up a base at Dahran. The oil relationship was also widened to bring in more US companies. Socal, which brought in Texaco in 1936, changed the name of its Saudi subsidiary from the Californian-Arabian Standard Oil Company to the Arabian-American Oil Company – and in 1948 brought in Exxon and Mobil as part-owners. This was also the year in which Aramco discovered Ghawar, the world's biggest oil field.

US presidential doctrines and the Gulf

Successive US presidents' foreign policy statements – which are termed "doctrines", as if they were papal pronouncements – tightened the links between America and the Gulf: between energy and national security.

The Truman Doctrine

The Truman Doctrine of 1947 promised US military help to all countries perceived to be threatened by Communism. It was targeted mainly at Greece, Turkey and Iran, which were thought most at risk from Soviet expansionism. But Saudi Arabia, which was feared to be a target of Soviet subversion, also received US military aid and training in 1951's Mutual Defence Assistance agreement between the two countries.

The Eisenhower Doctrine

The Eisenhower Doctrine of 1957 went further, promising that Washington would send combat troops, if necessary, to defend important allies of the US. The only place in the Middle East where President Eisenhower dispatched US troops was to Lebanon, in 1958. But he also sent more US military aid (though not any troops), to Saudi Arabia.

Furthermore, he improved relations with the Arab world by opposing the joint UK-French-Israeli invasion of the Suez canal zone; this remains the last time the US ever seriously sided against Israel.

The Nixon Doctrine

The Nixon Doctrine of 1969 effectively pledged that the US would henceforth seek to work via key regional allies and bolster them rather than intervene directly itself. The immediate application of this was in Southeast Asia, where Nixon wanted to aid local allies in their fight against Vietnamese communists.

But it also had a clear application in the Gulf, because, a year earlier, Britain had announced that, in a final retreat from empire, it would withdraw all its military forces "east of Suez" by 1971 – that is to say, chiefly from the Trucial states in the Gulf (which became the United Arab Emirates) as well as from Malaysia and Singapore. The Nixon Doctrine

Richard Nixon's good relationship with Mohammed Reza Pahlevi, Shah of Iran, spanned many decades. The pair are pictured here at a meeting in October 1969.

made clear that the US would not try to replace Britain directly as the policeman of the Gulf, but would instead bolster its local allies. So the trickle of US weapon sales to the Gulf became a flood – to Saudi Arabia in particular and, even more so, to the Shah of Iran.

One of Nixon's defence secretaries later said he had been under instructions from the president "to sell the Shah anything he wants, short of nuclear aircraft carriers". The Iranian part of the Nixon strategy collapsed in 1979 with the fall of the Shah and the arrival of Iran's Islamic revolution.

But within a year the US had committed itself more closely than ever to the defence of the Gulf. The Soviet invasion of Afghanistan in December 1979 brought Soviet influence within a few hundred kilometres of the Gulf and caused alarms bells to ring for Washington and its key Arab allies in the region, already made nervous by the Iranian revolution.

The Carter Doctrine

Within a month, in his January 1980 State of the Union speech, President Jimmy Carter proclaimed that "any attempt by an outside force to gain control of the Persian Gulf region will be regarded as an assault on the vital interests of the United States, and such an assault will be resisted by any means necessary, including military force".

To back his words, Carter established a Rapid Deployment Force, the precursor of Centcom (see box), capable of quickly reaching the Gulf in an emergency, and he set up several forward bases in the Gulf's vicinity.

The Reagan Doctrine

The Reagan Doctrine was a wider campaign waged by overt and covert means against Soviet and left-wing influence around the world. It was not particularly linked to the Gulf – or to oil – except that President Ronald Reagan enlisted the support of Saudi religious fundamentalism in a US-orchestrated campaign to oust the Soviets from Afghanistan.

Into this anti-Soviet movement, Washington co-opted Arab nationalists, a strategy that backfired horribly on the US in the form of the Taliban and al-Qaeda terrorists. Reagan also showed how important the free passage of oil to world markets was to the US when, in the later stages of the Iran-Iraq war of 1987–88, he allowed Kuwaiti oil tankers to fly the Stars and Stripes and gave them US naval protection from Iranian attacks. Reagan also successfully pushed through the sale of very sophisticated Airborne Warning and Control System (AWACS) aircraft to the Saudis.

Centcom – the US miltary's energy command

In the wake of the Soviet invasion of Afghanistan and the Iranian revolution, the US set up a new military command called Centcom, with its HQ in Florida. It's referred to as "central" because it occupies the central zone between the US commands in Europe and Asia. But it has also been central to US military activity in recent years.

Most of the US servicemen killed in combat since the mid-1980s have been under Centcom command. It has been involved in four wars – marginally in the Iran-Iraq war (protecting Kuwaiti tankers), but heavily in the two Iraq wars and the ongoing campaign, unrelated to oil, in Afghanistan. (After the 9/11 attacks, Centcom's responsibilities were extended to cooperation with central Asian states helping the US in the northern flank of Afghanistan.)

The Gulf War of 1991

Iraq's invasion of its neighbour Kuwait in August 1990 brought an immediate reaction from the US. Washington was all too aware that not only did Saddam Hussein now have Kuwaiti reserves at his disposal, but he also had the biggest Saudi oil fields in eastern Saudi Arabia within his striking distance. Shortly after the invasion, President George Bush senior made the oil factor explicit, stating: "our jobs, our way of life, our own freedom and the freedom of friendly countries around the world would all suffer if control of the world's great oil reserves fell into the hands of Saddam Hussein".

Saddam Hussein's naked aggression had put Iraq entirely in the wrong and made a United Nations Security Council resolution against Iraq a foregone conclusion. But had no oil been involved, it is unlikely that the US and nearly thirty other countries would have assembled forces of nearly one million men to eject Iraq out of Kuwait. (Bush senior was known to hate broccoli; he'd once joked that the reason he became president was so that nobody could order him to eat it. Others in the White House joked at the time of the Gulf War that "if Iraq had only a strategic stake in the world broccoli market, the US would not go near it".)

For more than a decade after the Gulf War, there was a policy of "containing" Saddam Hussein, enforcing UN resolutions that created "no-fly zones" in the south and north of Iraq, so that the Iraqi air force could not bomb Shia in the south or Kurds in the north.

During this period, the US strengthened its military presence in the region, building bases in Qatar, Bahrain and Kuwait. The cost of imposing no-fly zones, patrolling the waters of the Gulf, and providing training and equipment to the region's military totalled some $50–60bn a year. Keeping

America and the Gulf – a timeline

▶ **1971** Over-stretched Britain withdraws all its military forces east of Suez, including the Gulf, leaving a military vacuum that the US fills.

▶ **1973** The Arab-Israeli war breaks out, leading to the Arab oil embargo of the US. Fearful of the Soviet Union coming in on the Arab side, the US puts all its forces on alert and makes contingency plans to seize Saudi oil fields, if necessary to keep them out of Soviet hands.

▶ **1980** Alarmed by the Soviet invasion of Afghanistan posing a threat to the Gulf, President Jimmy Carter announces that "any attempt by an outside force to gain control of the Persian Gulf region will be regarded as an assault on the vital interests of the US, and such an assault will be resisted by any means necessary, including military force".

▶ **1980–88** The eight-year Iran-Iraq war breaks out in 1980. During the war's later stages, when Iranian forces start attacking Kuwaiti tankers in the Gulf, President Ronald Reagan authorises these tankers to be "reflagged" as US vessels and gives them US navy protection.

▶ **1991** After Kuwait refuses Iraqi demands for territorial concessions, including on cross-border oilfield, Iraq invades Kuwait in 1990. In response, the US leads the United Nations coalition in 1991 to oust Iraqi forces from Kuwait, and subsequently builds bases in Qatar, Bahrain and Kuwait.

▶ **2001** Involvement of a number of individual Saudis in the 9/11 attacks brings home to many Americans the exposure to terrorism and oil-related conflict that their country's close relationship with the Saudi government might be bringing.

▶ **2003** The US, leading its own coalition, invades and occupies Iraq. The US cites Iraq's likely possession of mass destruction weapons as its reason for its actions, but Iraq's oil is a background factor.

some US forces on in Saudi Arabia, despite the original promise to pull them all out after the 1991 Gulf War, may have fanned anti-Americanism there.

In general, during the 1990s, the oil issue that most concerned the US military was not that Saddam Hussein might attempt another direct grab at a neighbouring oil producer, but that, given his confirmed proclivity for aggression, he might use Iraq's oil wealth to build or buy weapons of mass destruction.

The Iraq invasion of 2003

Much ink has been spilt on the causes of the invasion. At the time of the 1991 Gulf War, Iraq was known to have possessed some chemical

US soldiers standing guard at the Ministry of Oil in Baghdad, in April 2003 – a few days after oil once more began to flow from Iraq's southern oil fields following the US-led invasion.

weapons. After the conflict, the UN had prohibited Iraq from developing any weapons of mass destruction and insisted on UN inspections to prove compliance. Saddam Hussein persistently frustrated the inspectors in their work, in what was an affront to the UN's authority. But this was nothing new: such behaviour had been tolerated for years.

What suddenly made this intolerable in Spring 2003? There are many plausible answers to this. Among them are President George W. Bush's personal wish to complete the job of ousting Saddam Hussein that his father left undone; the ideological desire of his neo-conservative entourage to force the pace of regime change in the Middle East; and last but not least, the misinformation supplied by Iraqi exiles and the gullibility of those in Washington who listened to them.

What is of most interest for the purposes of this book is whether oil was a factor, and if so, how big a factor. The administration of President George W. Bush had had energy on its mind right from the start. Vice president Dick Cheney presided over a task force that came up with a national energy plan in May 2001, within three months of taking office. Part of it focused on the domestic electricity sector, with sound analysis and prescriptions. Another part of the plan, which was focused on oil, was interpreted by many as the pay-off for oil-company funding of the Bush/Cheney campaign in 2000. Cheney, who had previously headed the Halliburton oil services company, brought some of this suspicion upon

Two masked protestors, one of US Vice President Dick Cheney, one of George W. Bush, during a protest in front of the Washington office of Kellogg Brown and Root Inc (a subsidiary of Halliburton), in February 2004. The Pentagon had recently anounced it had opened a criminal investigation against the company, with allegations of potential overpricing of fuel delivered to Baghdad.

himself by refusing to name whom he had consulted in drawing up his energy plan.

In fact, the Cheney energy plan's analysis of the problem of growing US reliance on oil imports was incontrovertible. The contentious parts, questioned by the plan's critics, were the proposed solutions. The Cheney plan almost totally ignored the demand side of the equation – the possibility of influencing the consumption of oil by taxation or regulation. This was in line with Cheney's remark, a few days before his report appeared, that "conservation may be a personal virtue, but it is not a sufficient basis for a sound, comprehensive energy policy". In terms of energy security, the Cheney plan was too attracted by the lure of diversification. It put too much faith in getting more oil from Western hemisphere neighbours and from the Caspian Sea, in the hope of buying less from the Middle East and the Gulf.

Within four months of the Cheney energy plan, the attacks of September 11 2001 happened. They immediately put everything in a new light, including the issues surrounding Middle Eastern oil. The attacks filled Americans with horror, and the fact that the majority of the perpetrators were Saudi citizens seemed evidence for many of how their country's close partnership with Saudi Arabia had backfired. It appeared more than ever to be a relationship of "oil for insecurity". As time went on, oil may have become a contributing reason for the invasion of Iraq: an Iraq free of UN sanctions and with some new foreign investment, would put more oil on to the world market. This in turn could help reduce Saudi Arabia's politi-

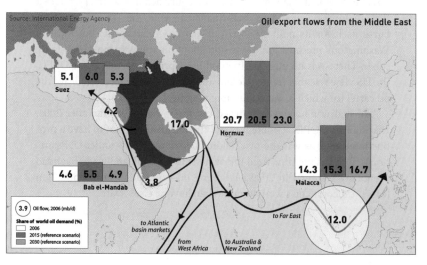

A small aircraft flies over the hijacked Saudi tanker MV Sirius Star on 9 January 2008. It is thought that the plane parachuted a container of ransom payment to the Somali pirates holding the vessel.

cal and economic leverage over the US and other oil consumers, which in the aftermath of 9/11 would be seen as welcome.

In some corners of the Bush administration, oil was clearly a more direct enticement. Paul Wolfowitz, the deputy defence secretary, forecast that an invasion would be "self-financing", with the US able to reimburse itself for its military costs out of Iraq's oil revenue. But most of Washington's pre-invasion discussion about oil seems to have been fairly sober. There was no push for Iraqi oil to be privatized and, as for OPEC, the US view was that Iraq could not and should not be forced to leave the oil cartel (of which it was a founding member).

And that was how events turned out in US-occupied Iraq after 2003. Iraq stayed in OPEC (where it is still the only country not to have a production quota, as was the case before 2003 when it was under UN sanctions). Iraq has not privatized the ownership of its oil, and when Baghdad eventually resumed in 2009 auctioning oil exploration and production leases, US companies won no more than anyone else's share.

Important world oil transit chokepoints

Name	2006 est. oil flow (millions of b/d)	Width at narrowest point	Oil source	Primary destination
Strait of Hormuz	16.5–17	21 miles	Persian Gulf Nations (incl. Saudi Arabia, Iran and UAE)	Japan, the US, Western Europe, other Asian countries
Strait of Malacca	15	1.7 miles	Persian Gulf Nations, West Africa	All Asia Pacific consumers (incl. Japan and China)
Suez Canal/ Sumed Pipeline	2.5	1000 feet	Persian Gulf Nations (especially Saudi Arabia), Asia	Europe and the US
Bab-el-Mandeb	3.3	18 miles	The Persian Gulf	Europe and the US
Turkish Straits	2.4	0.5 miles	Caspian Sea region	Western and Southern Europe
Panama Canal	0.5	110 feet	The US	The US and Central America

Source: US Department of Energy

The world's oil policeman

The US effectively acts as the policeman of global oil, keeping it flowing through the sea lanes of the world. This is not altruism. The blockage of any major oil route would raise the cost of oil and its biggest importer, the US, would be hit the hardest. Protection of sea lanes around the world is perhaps the principal justification for the US maintaining such a large navy – not only its Fifth Fleet, based at Bahrain inside the Gulf, but also other fleets that see it as their responsibility to help keep the Suez and Panama Canals, and the Malacca straits, open to shipping in general and tankers in particular.

Threats and alternative routes

Name	Past disturbances	Alternative routes
Strait of Hormuz	Sea mines were installed during the Iran-Iraq War in the 1980s. Terrorist threats post-September 11 2001.	East-West pipeline (745-mile long) through Saudi Arabia to the Red Sea.
Strait of Malacca	Disruptions from pirates are a constant threat. There was a terrorist attack in 2003. Collisions and oil spills are also a problem: there is poor visilibility due to smoke haze.	Rerouting through the Lombok or Sunda Strait in Indonesia. A pipeline could be constructed between Malaysia and Thailand.
Suez Canal/ Sumed Pipeline	Suez Canal was closed for eight years after the Six Day War in 1967. Two large oil tankers ran aground in 2007, suspending traffic.	Rerouting around the southern tip of Africa (the Cape of Good Hope: an additional 6000 miles.
Bab-el-Mendab	USS Cole was attacked in 2000, and a French oil tanker in 2002. Both were made off the cost at Aden, Yemen.	Northbound traffic can use the East-West oil pipeline through Saudi Arabia. Rerouting around the southern tip of Africa (the Cape of Good Hope: an additional 6000 miles.
Turkish Straits	There have been numerous past shipping accidents due to the strait's sinuous geography. Some terrorist threats were made after September 11 2001.	No clear alternative: potential pipelines have been discussed, incl. a 173-mile pipeline between Russia, Bulgaria and Greece.
Panama Canal	Suspected terrorist target	Rerouting around the Straits of Magellan, Cape Horn and Drake Passage: an additional 8000 miles.

Source: US Department of Energy

The flow of oil is becoming harder to police, because trade in oil is sharply on the rise. There is a growing geographical mismatch between where demand is growing fastest (mainly China and India) and where the potential for more supply is greatest (the Middle East). As a result, more oil than ever is on the move. The oil trade between the major regions of the world is expected by the IEA to be 55m b/d in 2030 (more than half world oil output) or 35 percent more than today. This trade is carried by some 4000 tankers plying the world's oceans, to which must be added a growing number of Liquefied Natural Gas (LNG) tankers.

All tankers are slow. Some are also very cumbersome to manoeuvre, because of their size. "Super-tankers" are more than 500,000 deadweight tonnes. So they are vulnerable to attack either on the high seas or as they have to pass through narrow chokepoints. A very high proportion of the world's tankers have to pass through the Strait of Hormuz to exit the Gulf, and then – depending on whether they are west- or east-bound – either the Bab-el-Mandeb entrance to the Red Sea and the Suez Canal or the Straits of Malacca between Indonesia and Malaysia.

So far, there have only been a few attacks on oil tankers (and none yet on an LNG tanker). The attacks have been limited to a few incidents inside the Gulf during the Iran-Iraq war of the 1980s, an al-Qaeda attack on a French tanker off Yemen, and the capture of a Saudi tanker in the Red Sea by Somali pirates in 2008. Nonetheless the impact on world oil trade of one or other chokepoints being closed off would be considerable.

US energy independence
motivator and mirage

In the light of all the security entanglements that oil dependence has brought the US, it is not surprising that many Americans and their presidents have worried for decades about increasing US reliance on oil imports. These constituted 30 percent of total oil consumption at the time of the 1973 Arab oil embargo on the US, and amount to 65 percent today. Even though Americans depend on foreign oil no more than Europeans, and far less than Asians, they fret more than others about oil import dependence. There are several reasons for this. At the time of World War II, the US was the premier oil *exporter*: its current status is something of an Achilles heel for a world superpower. It also links Washington to some very conservative Arab regimes, and in the context of the Arab-Israeli conflict, could hypothetically compromise the US's support for Israel, (though there is no evidence of this in practice).

So calls for energy or oil independence have become a mantra of successive presidents: Nixon, Ford, Carter, Reagan, George W.Bush, and now Barack Obama. It's a theme that resonates well with US public opinion – it's more persuasive than climate change as a motivation for practising energy conservation and promoting alternative energy. Drives for energy independence have been used to justify measures ranging from the introduction of vehicle fuel-economy standards and a lower national speed-limit in the 1970s to the development of ethanol road fuel and renewable electricity in the 2000s.

The fact that these measures have not prevented the ratio of US oil imports to home-grown energy production from rising is not evidence of total failure: without these measures, the oil-import ratio might have gone higher still. But actual energy independence for the US is a mirage, a goal unlikely ever to be reached.

Drill, baby, drill!

This was a slogan that emerged, on the Republican side, in the 2008 US presidential election, as a call for restrictions on domestic drilling to be eased. For the US is probably the only country in the world to have put certain of its hydrocarbon reserves off-limits to commercial drillers. Alaska has two areas off-limits to drillers. The National Petroleum

Drill baby drill? Caribou graze in the Arctic National Wilderness Refuge: a conservation area off-limits to oil and gas exploration.

Reserve is an area originally set up to maintain a reserve supply for its navy, while the Arctic National Wilderness Refuge is, as its name suggests, a wilderness reserve.

Moreover, at the insistence of several environmentally- conscious coastal states, there has been basically no drilling off either the US east or west coast, nor in the eastern part of the Gulf of Mexico along the Florida coastline. If all these restrictions were lifted, it is probable that the US could, at least briefly, reverse the post-1970 peak in its oil output. After the massive BP oil spill in the Gulf of Mexico in 2010, the lifting of restrictions elsewhere seems unlikely.

But as Anthony Cordesman, an energy and defence expert with CSIS in Washington, points out: "it is not clear that you would solve any strategic problem by depleting US resources first [ahead of foreign producers]". The Obama administration wants to lower the import share of oil and oil products to below 60 percent (from 65 percent today), but Cordesman does not believe it will ever go below the 50 percent mark.

Changing oil suppliers?

There is no good oil or bad oil – just oil. Both George W. Bush and Barack Obama have implied that they would like the US to switch suppliers in calling for the US to cut imports by an amount equivalent to its oil purchases from the Middle East. In fact the US buys relatively little directly from the Middle East. The top five oil suppliers to the US are, in descending order, Canada, Mexico, Saudi Arabia, Venezuela and Nigeria.

But switching suppliers would not help, says Jim Woolsey, a former director of the Central Intelligence Agency who believes passionately in the potential of electric cars to break the US oil habit. "One changes nothing by changing the pattern of oil purchases – if the US were to buy only from Norway it would affect nothing because the US would still be buying in a world oil market that is subject to the influence of OPEC and the Saudis," Woolsey maintains.

Moreover, all America's trading partners and allies in Europe and Asia would still be in that world oil market. The only real energy independence would be zero oil imports from anywhere.

Future frictions

and Arctic angst

Climate change will multiply existing environmental, economic and geopolitical problems. It will aggravate the problems of desertification and water scarcity that already afflict the broad belt that runs around the earth's equator – through Africa, the Middle East, Asia and Central America. But it will also create geopolitical tensions over energy in a new area – the high north of the Arctic Circle.

For the melting of the Arctic sea ice brings "new opportunities and new risks" in the words of Per Stig Muller, Foreign Minister of Denmark, an Arctic coastal state via its possession of Greenland. These opportunities are mainly economic, resulting from the possibility of exploiting the considerable reserves of gas and oil that are reckoned to lie within the Arctic Circle. Russia is developing its giant Shtokman gas field in the Barents Sea, and Greenland is already licensing explorations blocks off its west coast and will shortly be doing the same off its east coast to the likes of Cairn Energy of the UK. These ventures risk exacerbating international political tensions over rival energy-resource claims, and present environmental dangers of pollution from energy development as well as from increased shipping.

It's clear that Arctic energy will not remain largely unexplored, uncharted and unclaimed for much longer. The Arctic is more immediately vulnerable than Antarctica to global warming, because it is warmer. It is basically a sea surrounded by continents, with the North Pole appreciably warmer on average than the South Pole in the middle of Antarctica, a continent surrounded by seas. The Arctic is also more vulnerable to exploitation than the Antarctic, which is protected by international treaties from territorial claims and commercial development. The main reason for this discrepancy in the protection of earth's two poles lay in the Cold War. The two rival superpowers of that time, the US and the Soviet Union, agreed the 1959 Antarctic treaty, which allows freedom of scientific research on the cold continent but nothing else, and bans all military activity there.

The reason for this is that neither side had the slightest interest in being able to fire missiles at each other "the long way round" – via the South Pole – when they could strike each other far more quickly over the shorter North Pole route. This, plus the greater uncertainty of legislating on an area essentially consisting of floating ice, explains why governments never reached any specific agreement on the status of the Arctic.

Arctic energy resources

Of the six percent of the earth surface encompassed by the Arctic Circle, one-third is above sea level, and another third is in continental shelves

Undiscovered gas (trillion cubic feet)

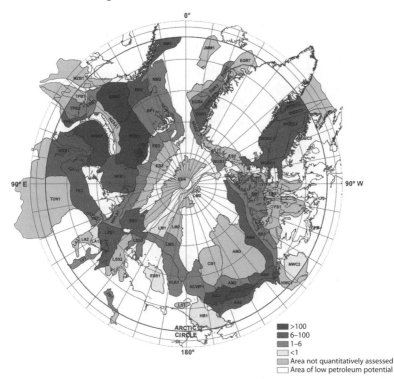

- ■ >100
- ■ 6–100
- ■ 1–6
- □ <1
- Area not quantitatively assessed
- Area of low petroleum potential

beneath less than five hundred metres of water, with the remaining third being in deep ocean that is considered to have little hydrocarbon potential. Some of the one-third above water has already been explored and exploited, notably America's Alaska North Slope and Russia's west Siberian basin. And these drillings have contributed to estimates of what may lie undiscovered in the rest of the Arctic region.

In a report released in *Science* magazine in May 2009, the US Geological Survey estimated that about 30 percent of the world's undiscovered gas and 13 percent of the world's undiscovered oil may lie within the Arctic Circle. Most of these resources are believed to be offshore, but in depths of less than five hundred metres of water. This ought to make them commercially viable (though this was not investigated by the USGS survey, which takes no account of technical or economic factors).

Arctic oil (billion barrels)

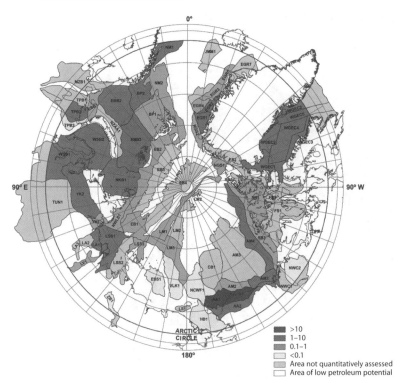

The estimated Arctic oil is probably not significant enough to shift the geographic pattern of world oil production, which is centred on the Middle East. But the amount of undiscovered gas is three times greater, measured on an energy-equivalent basis, than the oil, and is largely concentrated in Russia or in Russian territorial waters. This would confirm the long-term dominance of gas by Russia, which already holds 25 percent of proven gas resources (a higher ratio that Saudi Arabia's 21 percent share of proven oil reserves).

This wealth has not inhibited Russia trying to further increase its share of potential Arctic hydrocarbons. In 2007 one of its submarines planted a flag underneath the North Pole, to underline Russia's claim that a large part of the Arctic seabed, the Lomonosov Ridge, is just an extension of the Russian continental shelf. Moscow's reason for claiming such an extension relates to the United Nations Law of the Sea treaty, which came into

The Russian research vessel *Akadmik Fyodorov*, on sail in the Arctic Ocean. The ship carries miniature submarines – one of which was used to plant a Russian flag on the seabed beneath the ice of the North Pole.

effect in 1994. This gives a state an Exclusive Economic Zone over sea that is within two hundred nautical miles of its coast – and even further if it can show its continental shelf extends beyond the two hundred-mile limit. The four other Arctic coastal states with territory inside the Arctic circle – Canada, the US, Norway and Denmark – have not accepted the Russian claim. These four are all NATO members. So there is scope for serious geopolitical tension here.

Russia is impossible to ignore in the Arctic. Its territory covers almost 180 degrees of the Arctic circle. Encouragingly, Russia joined the other four coastal states in a declaration at a meeting in Ilulissat in Greenland. In their Ilulissat declaration, the five said they would settle all jurisdictional and navigational issues in the Arctic ocean themselves within the framework of the Law of the Sea treaty, without the need for any new international legal regime.

The US is not yet in a position to use the Law of the Sea treaty to extend its jurisdiction, because unlike the four other Arctic coastal states its Senate has failed (as with many other international treaties) to ratify it.

Arctic waterways

The melting of the Arctic sea ice will make routes as well as resources contentious. With global warming, the dream, which goes back to sixteenth-century English navigators, of an ice-free a Northwest passage, between the Atlantic and Pacific will be realized: it is thought that the shrinkage of sea ice will mean the passage will be open for much longer passages of time. But Canada finds the idea that a passage through its Arctic islands should be considered an international waterway – as the US insists – completely anathema. A shorter, northern shipping route between Europe and Asia would reduce energy consumption and produce fewer emissions. But this benefit would be outweighed by the environmental damage from shipping and any oil and gas drilling in a very delicate environment – not to mention the inevitable political tensions that would result from states scrambling to carve up Arctic energy assets. In July 2007, Canada's prime minister, Stephen Harper, made an assertive statement, in announcing the establishment of a new deep-water port in the far North. "Canada has a choice when it comes to defending our sovereignty over the Arctic," he said. "We either use it or lose it." Ottawa has reversed the rundown of its navy so as to be able to police its Arctic waters properly.

PART 5
ENERGY
PROSPECTS

What will the future look like?

Saving energy
Getting better all the time

The first parts of this book stressed our absolute need, in order to try to preserve a habitable planet, to move to a low-carbon economy. We have surveyed today's energy landscape, the range of fossil fuels and renewable energies and highlighted some of the existing drawbacks – many of which have nothing to do with climate change – of living in a fossil-fuelled world. So now we arrive at the prospects for change. It's important to look at why the world has shown such inertia in changing our energy system, at what progress we have made in using less energy, and at the remaining barriers we face in saving energy.

Given that we know we're travelling along the wrong energy path, the prospects for changing direction ought to be better than they are. But climate change is a uniquely difficult problem for many reasons. Among them are the fact that the costs and benefits of tackling (and of not tackling) it are unevenly spread – both across time and between countries.

The obstacles

In terms of time, it's today's generation that will have to pay the cost of an emissions reduction that will largely benefit only future generations. The four- or five-year electoral cycle in democratic countries is far too short to take this into consideration, favouring policies that will appeal to voters making decisions in the here and now.

There is also a geographical mismatch. Generally, the industrialized countries most responsible for the build-up of greenhouse gases are not those feeling the first effects of climate change, which have mainly been felt in poorer developing countries. Americans, for instance, might be more motivated to cut their high-level emissions if they were as generally threatened by rising sea levels as Bangladeshis are. Rich nations are also better able to defend themselves against extreme weather conditions that afflict them. But there are additional, specific reasons for inertia when it comes to changing energy systems.

Slow turnover

The longevity of energy-hungry commodities and buildings provides few opportunities for replacement with more efficient equipment or buildings. According to some averages worked out by the International Energy Agency, housing stocks last anywhere between 40 and 400 years, industrial buildings between 10 and 150 years, large hydropower plants between 60 and 120 years, coal-fired plants from 40 to 60 years, nuclear reactors over 40 years, and power grids and gas pipelines around 40 years. At the top end of the product range are aircraft, the design of which can remain unchanged for more than fifty years (from first studies on the drawing board to the last plane coming off the production line).

According to the UK's Energy Saving Trust, the average life of a fridge is 12.8 years, and that of a freezer between 15 and 17 years. The only electrical item we replace very often is Thomas Edison's incandescent (and inefficient, as well as short-lived) light bulb. Of course, sometimes there arises either the necessity to accelerate this natural cycle of replacement, as in the case of repairing damage after a war, or the opportunity to accelerate it, as in the case of the world-wide economic recession of 2008–09.

Habits and lifestyle

Reluctance to change personal behaviour is another factor. Some energy-saving measures do not require any behavioural change. Installing a more efficient boiler or buying a hybrid (part-electric, part-petrol) car does not involve living in a colder house or travelling less. Turning the heating thermostat down in winter or the air-conditioning thermostat up in summer merely requires the minor sacrifice of adding or subtracting a layer of clothing. But any restriction on personal travel tends to be regarded as a major sacrifice. This is why people react with panic to any form of petrol rationing or shortages.

Another aspect of the mobility revolution is air travel, which budget airlines have made affordable to many people. One example of the particular store that even environmentalists set on maintaining their mobility was brought to light by a 2007 survey by the British Market Research Bureau. This surveyed nearly fifty thousand people, over two years, by asking

You can't beat an oil shock for energy saving

It didn't seem like it at the time – with long queues at petrol stations on both sides of the Atlantic and a rota system for road-use according to whether the vehicle licence plate ended in an odd or even number. But the quadrupling of the oil price between October 1973 and 1974, and the subsequent price surge in 1979–80 was a tremendous help to energy conservation.

All the big improvements in energy efficiency came in the aftermath of 1973. It is a salutary fact that the 1970s oil-price shocks did more to cut energy use and carbon emissions than all the policy changes of the 1990s and 2000s have produced.

The industrialized oil-consuming countries that belong to the International Energy Agency improved their energy efficiency by 2 percent a year from 1973 to 1990, but by only 0.9 percent a year from 1990 to 2004. Some governments took the 1970s oil-price shock seriously enough to embark on radical policy action: France went for nuclear power; Japan redoubled its energy-efficiency efforts; and Denmark focused on saving energy and on developing a new alternative source with wind power. (Almost twenty percent of Denmark's electricity was produced by wind power in 2007.) Other governments simply let the market mechanism of higher oil prices do the conservation for them.

The post-1973 efficiency improvements produced a lasting gain. In a study of eleven of its larger member countries, the IEA concluded that "without the energy efficiency improvements achieved since 1973, energy use for the eleven IEA countries would have been 56 percent higher than it actually was …this makes energy savings the most important "fuel" for the IEA-11 in this time period."

The amount of energy saved in 2004 was slightly more than the oil that these eleven countries consumed. This concept of the quantity of energy *not* used being considered the most valuable fuel of all is interesting, and logical, because it involves building nothing and emitting nothing. This non-energy is sometimes referred to as "negajoules" or "negawatts" – words that look like typos but are plays on the energy measurements of megajoules or megawatts.

As time wore on and the shocks wore off, most countries grew lax about energy conservation. The same laxness occurred in energy research and development. When the oil price came down in the mid-1980s (and stayed down thereafter for more than fifteen years), so did energy research and development. It has only recently picked up.

If, for instance, the 1980 peak in energy research had been sustained, the European Union and its member states would be collectively spending €7-8bn a year on energy, instead of around €2.5bn a year. Fifteen years of fairly low oil prices, from 1986 to the turn of this century, left the EU as a whole, in the words of the European Commission, "with accumulated under-investment [in energy R&D] due to cheap oil".

them to fill in a questionnaire with fifteen measures of consumer behaviour, such as "green" product purchases and subscriptions and donations to "green" groups. Those exhibiting five or more these fifteen lifestyle traits were rated as "Active Environmentalists". Over ninety percent of this group said they were ready to pay more for environmentally friendly products, said they were worried about traffic pollution and congestion, and said people had a duty to recycle. Yet the majority of these Active Environmentalists, who were rated as well educated, affluent and middle class (ABC1), and were aged 35 or over, were found to be likely to fly three or more times a year. This is more than the average UK citizen.

The survey (entitled "Can Fly, Will Fly") found that the Active Environmentalists were no less likely to fly than the ABC1 group in the population as a whole. It concluded that "all the good intentions of the Active Environmentalists are neutralized by their love of travel". The participants had not been asked whether they had bought into carbon-absorbing schemes offsetting their air travel emissions. (Many of which have anyway been derided as inadequate by staunch environmentalists.) It underlines what we all know – adopting a "low carb" energy diet is easier said than done.

When efficiency doesn't equal saving

Energy efficiency can have a perverse effect on energy demand – the more energy you save, the more you can afford to use for something else; and the more efficiently energy can be produced and the cheaper it becomes, the greater the incentive to use more of

William Stanley Jevons gave his name to a paradox: that increases in efficiency can be the catalyst for increases in consumption.

it. This perversity is known as the "rebound effect": demand rebounds to nullify the efficiency improvement. The degree of rebound is usually expressed as a percentage of the estimated savings from an efficiency improvement. In other words, a rebound of twenty percent still leaves eighty percent of the expected savings as a net reduction in demand.

The rebound effect has been recognized ever since the nineteenth-century invention of the coal-fired steam engine. In the book he published in 1865 called *The Coal Question*, William Stanley Jevons observed that coal consumption had soared since the introduction of James Watt's coal-fired steam engine, because this made coal a more cost-effective power source. Jevons argued that any further efficiency improvements in the use of coal would increase the consumption of coal.

This reasoning has sometimes been termed the "Jevons Paradox", and used to claim that all energy efficiency moves are therefore futile. Energy economists used the term "backfire" to describe this counter-productive effect, which also occurred after the introduction of the electric motor in the twentieth century. Energy efficiency improvements can certainly stimulate demand so much that overall energy use actually increases – in other words the rebound is more than one hundred percent of the expected energy savings.

However, this ignores the fact that increased welfare and income can come from efficiency improvements, and also that the energy appetites of individuals (though not necessarily companies) are ultimately limited by their personal needs. Some rebound of demand is probably inevitable with any genuine improvement in energy efficiency. A 2007 study by the UK Energy Research Council (UKERC) estimated the rebound effect in household energy and personal transport to be less than thirty percent and nearer ten percent.

With these rebound and backfire concepts in mind, one should always exercise a certain caution in estimating the actual energy savings from energy efficiency technology. Looking forward, the backfire effect is unlikely to occur again – not for lack of revolutionary advances, but because so many extra charges (such as the costs of carbon permits and of renewable energy subsidies) are being loaded on to energy bills that energy is not likely to get any cheaper.

But for a fascinating view of past rebound and backfire effects, take a look at the table overleaf, cited in the UKERC study. The amount of lighting that the average individual was using increased much faster than their income. This was because of the precipitous fall in the price of lighting services, which was in turn the result of the huge increase in lighting efficiency and reductions in the cost of fuel for lighting. On these figures it would seem that, over the past two centuries in Britain, if lighting efficiency improved 1000 times but per capita consumption of lighting 6566 times, there has been a six-fold increase in average consumption of energy for lighting.

Seven centuries of lighting in the UK
(year of 1800 = 1.0 for all indices)

	1300	1700	1750	1800	1850	1900	1950	2000
Price of lighting fuel	1.5	1.50	1.65	1.0	0.40	0.26	0.40	0.18
Lighting efficiency	0.5	0.75	0.79	1.0	4.4	14.5	340	1000
Price of lighting services	3.0	2.0	2.1	1.0	0.27	0.042	0.002	0.0003
Consumption of light per capita	–	0.17	0.22	1.0	3.9	84.7	1528	6566
Total consumption of light	–	0.1	0.15	1.0	7	220	5000	25,630
Real income per head	0.25	0.75	0.83	1.0	1.17	2.9	3.92	15

Source: Roger Fouquet and Peter Pearson, *Seven Centuries of Energy Services: The Price and Use of Light in the United Kingdom (1300–2000)*, 2006

Getting more out of less?

In general, developed economies are getting better at using less to make more. Surveying 19 of its members, the IEA found that between 1990 and 2004, total energy use in manufacturing increased by only 3 percent, while the value of their manufactured output rose by 31 percent. Since as a group they were also using cleaner fuels – with the share of gas (and electricity) growing slightly at the expense of oil and coal – the IEA countries' emissions only increased by one percent.

This decoupling of economic growth from energy and CO_2 emissions is just what we want to see. It is particularly satisfying that the most energy-greedy sectors, such as steel, minerals, chemicals and paper and pulp, have all cut their energy use, with chemicals achieving the largest reduction of nearly thirty percent. But some countries' reduced energy-usage has had less to do with any breakthrough in efficiency or technique and more to do with their shift away from heavy industry or raw material production. For instance, the reason why Australia has the highest energy intensity of any IEA member country is that half of its industrial output cies in the production of raw materials.

With the maturing of the IEA's nineteen richer economies, the service sector (which covers such a vast range of white-collar activities that it can hardly be called a sector, ranging from trade, finance, real estate to hair cutting) has gained at the expense of manufacturing. Between 1990 and 2004, service output rose by 45 percent, but energy consumption rose by only 26 percent. Not surprisingly it is electricity – versatile and clean (at the point of use) – that dominates the service sector(s), because of the electronic equipment and air conditioning in offices.

In households, total energy use is on the rise – it rose by fourteen percent between 1990 and 2004 among these IEA member countries. This is because the population is also growing and because more of us now live in smaller household units. But the new growth factor is the plethora of appliances that now litter our houses.

It is not so much that individual appliances are so wasteful. Indeed all of the traditional household appliances use energy more efficiently than they used to: the only exception being televisions, as today's large screens use more power. The five big appliances that most people have – fridges, freezers, TVs, washing machines and dishwashers – still account for about half the energy consumption of all household appliances. But this ratio has been declining as smaller gadgets proliferate and push up overall electricity use.

Gadgets galore

We can use information technology and electronic devices to save energy by, for instance, creating smart grids and by teleworking from home instead of travelling to offices. But we are also in danger of letting these technologies gobble up a serious and growing amount of energy, often heedlessly and needlessly. Charting the problem in a 2009 report called *Gadgets and Gigawatts*, the IEA estimated that information and com-

In late 2009, this was the largest plasma screen TV panel commercially available: a whopping 103 inches.

munication technologies (ICT) and consumer electronics (CE) currently account for around fifteen percent of global residential electricity consumption – and that this would double by 2022 and triple by 2030. By this last date the bill for this electricity use could be $200bn, or the current residential electricity consumption of the US and Japan combined.

Growing global ownership of consumer electronics is, in principle, a good thing if it improves people's welfare and quality of life around the world. Many traditional electronic items are now cheaper to buy and cheaper to run. New refrigerators use much less power than the old models. But new scales are being found for traditional items. The liquid crystal display (LCD) TV screens may use less energy than the cathode ray tube monitors they are replacing, but they don't if they are twice the size. Of course, large-screen TVs are not just used for television broadcasts, but to display video games and DVDs. So it is not surprising that TV sales around the world are actually growing faster than the number of households gaining access to electricity.

Around one billion people now use a computer. Faster access to the Internet through broadband is high on the political agenda of many governments. Broadband makes it quicker to get a given piece of information off the Internet – but also makes it more tempting to stay on the Internet longer. New uses for the net proliferate – transmitting music, social networking, free telephony (skyping) and processing digital photographs – and so inevitably do the hours spent online and plugged in.

But what seems particularly wasteful is the number of gadgets left on standby or in sleep mode. It is a mistake to think that standby mode powers nothing but a little red light. Many devices consume between half and two thirds of their full power on standby. The European Commission estimated that in 2005 the total of 3.7bn household and office computers and electronic gadgets in use among the 25 countries then in the European Union used 47 terawatt hours (TWhs) of electricity while they were in standby mode. This standby consumption – roughly equivalent to the total annual electricity consumption of Greece – cost €6.4bn and caused 19m tonnes of carbon dioxide emissions. A survey in the UK at around the same time by the Energy Saving Trust found that the average household has up to twelve gadgets on standby (or charging) at any one time.

Some products require constant power to keep their time clocks running. TV set-top boxes, which are becoming more common as Europe moves to digital TV, need to have power in order to download information from digital transmissions which update their electronic programme guide and software. But in most cases the standby mode is pointless electricity consumption, designed at most to save people a few seconds delay when they start their computers or gadgets up. At this stage, it may be impossible to wave goodbye to standby. But products could at least be required to "power down" automatically to standby mode, and standby could be set at as low a level as possible.

Transport Trauma

Reinventing the wheel

The toughest decarbonization challenge lies in the transport sector. Transport is almost totally dependent (95 percent) on oil. This reliance on just one kind of fossil fuel means there is an inherent energy-security argument, as well as a climate-change argument, for saving on oil use in transport. But so far rail, whose passenger load is minimal on a global scale, is the only transport mode for which an alternative fuel – electricity – has been found. Around a quarter of the world's rail network is said to be electrified.

Change in the rest of the transport sector could be traumatic. To explain why, it is worth reflecting for a moment on the general nature of energy. It is a derived demand – we don't eat, drink, sit on or wear coal, gas, oil, electricity in their raw state. It is the energy services they provide – heating, cooling, cooking or locomotion – that we require. As consumers, we are essentially indifferent to the energy form that delivers the service. Most of the time, the form it takes is electricity, and most of us just think of electricity as something that comes out of the wall, with little concern for how it has been generated. We might prefer our electricity to be generated renewably with a wind turbine or a solar source, rather than in a nuclear reactor or by a coal-powered steam turbine. But all these sources produce the electrons, and as far as the stationary energy services of heating, cooling and cooking are concerned, one electron is as good as another.

Locomotion, however, is different. We all know how tightly all modern transport (bar rail) is locked into oil, and very many of us fear that a move away from oil might rob us of our mobility. We would not be indifferent to this. People around the world have become accustomed to the perception of freedom that mass-produced cars and mass air-travel has brought

Motorizing America

Naturally, the US oil industry had an interest in pushing the expansion of the car industry. The advent of the car was unreservedly welcomed in America's sparsely populated rural areas, where it was genuinely liberating for people. But according to the account in Ian Rutledge's book, *Addicted to Oil: America's Relentless Drive for Energy Security*, by the 1930s US car manufacturers were fretting at the sales resistance they were meeting in US cities where public transport-systems already existed in the form of electric streetcars. The president of the Studebaker Corporation admitted in 1934 that many well-to-do people did not own cars, not because they could not afford them, but because "as they will tell you, the ownership lacks advantage. They can use mass transportation more conveniently for many of their movements."

Coincidentally or not, some car and oil companies, notably General Motors, Chevron and Phillips Petroleum, started in the mid 1930s to buy up some streetcar and trolleybus companies, especially in California, and to convert them to the motor buses they were building or fuelling. Rutledge says: "It has been convincingly argued by a number of US transportation historians that the real intention was to clear the way for the mass introduction of the automobile in urban areas."

The reasoning for this claim is that operating a motor-bus service was less profitable and less reliable than the existing electric trolleybus and streetcar services. But, the argument goes, the unsatisfactory nature of the new motor-bus services actually suited the car and oil companies' purpose, because as the quality of urban public transport fell, so demand for private cars rose. Whether you buy this theory or not, the fact is that in the US of 1922 there were only 1370 miles of urban motor-bus routes in service compared to 28,906 miles of streetcar railway track. But by 1940 there were 78,900 miles of motor-bus routes, and operational streetcar track had fallen to 15,163 miles.

However, the federal authorities did play a direct role in motorizing America with the passage of the 1956 Federal Highway Act. This appropriated $21bn, a huge amount of money at the time, to build 41,000 miles of interstate highways over a 20-year period. Since then, this road network has been expanded and maintained from the proceeds of the federal gasoline tax. This federal tax is only 18.4 cents per gallon there are usually state gasoline taxes that are slightly higher, but most of its proceeds are dedicated to the Federal Highway Trust Fund, which finances the interstate road network.

What is unusual about the US is not the decrease during the twentieth century of electric public-transport systems – this happened in many European cities (though the disappearance of surface electric transport was more often compensated by subway systems in Europe than in the US). Nor has the US been unique in building such a national road network – think of the autobahns Germany built in the 1930s or the autoroutes France built more recently. What is unusual about the US is that it has devoted so much of so small a gasoline tax to subsidize the infrastructure, and thus encourage more driving.

them, even while growing road and airport congestion often make a mockery of the idea.

People's appetite for travel seems to be the hardest to assuage. If travel becomes cheaper, as air travel has with budget airlines, then people make more flights and take more holidays abroad. If travel becomes quicker, as it has with high-speed trains in Europe, then people travel further or more frequently. Most enticing of all is the car, with all its prized autonomy and flexibility.

The need for travel is obviously greater in large countries with spread-out populations. Yet it seems that income plays a more important role than distance. For instance, according to the OECD, the average American travels 30,000 km per year (all forms of transport combined), which is almost twice the annual travel of the average Canadian or Australian, whose countries are no more compact than the US. It is the higher income of the average American which makes the difference.

Multiply rising incomes with big population numbers, and you end up with environmentally frightening prospects, such as China's fleet rising from around 25m cars today to around 250m cars (roughly the number now in the US) by 2030. Happily, the Chinese government is also a bit concerned about this. It has introduced fuel efficiency standards that are tighter than the US's, as well as investing in electric cars and developing bio-fuels.

Some savings have been made on the fossil fuels going into transport. Among the majority group of IEA member countries, in the 1990–2004 period the distance travelled by their citizens rose 31 percent, while the energy used for transport (all modes) in these countries increased by only 25 percent. Tighter regulatory pressure on fuel efficiency in Europe and Japan, plus a shift to more efficient diesel engines in Europe, contributed to this relative saving.

More recently, the Obama administration has backed a tightening of America's Corporate Average Fuel Efficiency (CAFE) standards. Oil consumption in richer countries is flattening out. Tony Hayward, the chief executive of BP (which is a major player in the US as the result of its acquisitions there), said recently that "BP is unlikely to sell more gasoline to Americans than it sold in the first half of 2008" – before the 2008-09 recession hit, that is.

But oil consumption is still galloping ahead in developing countries. And there is only so much that tinkering with the internal combustion engine can do to reduce fossil-fuel consumption and the concomitant CO_2 emissions.

Road transport

Electric cars

It is beginning to look as though the only real answer, on energy security and climate-change grounds, to transport's dependence on oil is the electric battery-powered car. (Provided, of course, that the electricity used to charge them can be provided by low-carbon means.) These have been around for a long time. Indeed at the outset of the automobile age, in the years 1899 and 1900, electric cars outsold their two competitors in the US: cars driven by steam (quite impractical); and cars using the gasoline-powered internal combustion engine (the clear eventual winner over the subsequent century).

Ironically, electric-car technology undermined itself with the invention of the electric starter in petrol-powered vehicles. As Gary Kendall notes in his book *Plugged In* (written for the World Wildlife Fund) this "eliminated the need for the hand-crank and thereby neutralized a unique selling point that electric vehicles could previously claim: ease of use, especially for female drivers unwilling to turn the physically demanding crank". Kendall goes on to suggest that this "encouraged automotive battery manufacturers to focus on mass production of small, low capacity auxiliary batteries rather than on increasing storage capacity, which would have benefited the range of electric vehicles". His implication is that somehow electric batteries could have replaced the internal combustion engine earlier.

Li-ion

Forty, thirty or even ten years ago, such a shift is hard to imagine. For it has taken years of development in consumer electronics, where the energy-to-weight ratio of a battery is vital, to produce the lithium-ion battery that is now powering electric cars. There will probably always be a place for oil, probably in the form of diesel, to power trucks; it seems unlikely that electric batteries will ever hold enough power to use in long-distance road haulage. But with the advent of lithium-ion batteries, it looks as though electric vehicles will no longer be confined to the niche applications of milk floats, golf carts, forklift trucks and airport buggies.

Lithium-battery development began some thirty years ago at the Exxon oil company, an ironic birthplace were electricity to eventually displace petrol – though then, lithium batteries were never considered an alternative to convenient and cheap petrol. Repeated refinements by the consumer electronics industry have given lithium-ion (sometimes abbreviated to li-ion) batteries greater range and power. So far these batteries are made almost entirely in Asia (China and Japan), but part of the Obama administration's green-energy plan is dedicated to jump-starting li-ion battery manufacturing in the US. (Similar geographical diversification would also be desirable in another commodity – rare earth elements, which are much used in electrical equipment. China dominates world mining of rare earth elements, which are used to make permanent magnets, among other things, and it is increasingly using more of its own output of these elements and therefore reducing exports.)

Road range

The electric-car driver has a natural fear of running out of juice. Range is an issue, although surveys in the US, for instance, show that eight out of ten people commute fewer than forty miles a day. To deal with the range issue, there are three alternatives.

▶ Supplement the electric battery with an internal combustion engine

A lithium-ion battery pack composed of 6831 individual cells powers the Tesla Roadster, an all-electric sports car which has a 220-mile driving range and uses rooftop solar collectors to provide some of the energy to charge the car.

in a hybrid car. The hybrid concept has existed for decades, but it took the launch of the Toyota Prius in 1997 and the Honda Insight in 1999 to make it commercially viable (albeit at quite a high price). These and other hybrids have higher mileage and lower emissions than those powered by fossil fuel alone.

▶ Allow the electric battery in a hybrid car to be recharged by plugging into any electric source along the road. This is, unsurprisingly, known as a plug-in hybrid vehicle. GM is launching the first mass-produced model of this kind with the Chevrolet Volt in 2010. It is not clear how long this recharging would take.

▶ Set up an infrastructure that allows all-electric cars (such as the Nissan Leaf) to easily and quickly replace their batteries just as simply as if filling up with petrol. At present no such network of battery replacements exists. But Shai Agassi, the Californian creator of a company called Better

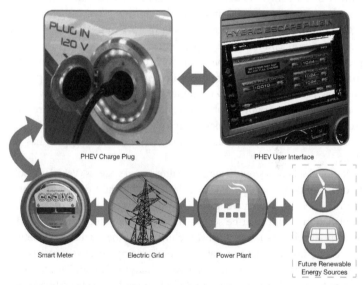

PHEV Charge Plug   PHEV User Interface

Smart Meter   Electric Grid   Power Plant   Future Renewable Energy Sources

In 2009, Ford unveiled its Intelligent Vehicle to Grid Communication System, which allows its plug-in hybrid electric vehicles to "talk" directly with the nation's electric grid. A driver can thus programme when to recharge the vehicle, for how long and at what utility rate.

For example, a vehicle owner could choose to accept a charge only during off-peak hours, when electricity rates are cheaper. Above is Ford's diagram showing how it fits into the energy supply chain.

Place, is aiming to create such networks around the globe. According to his plan (called Better Plan), the procedure for the battery exchange is that the car is driven over a pit, in which there is a robotic device that extracts the exhausted battery and inserts a fresh one – all in a couple of minutes, or the time it would take to fill up your car with petrol.

Road pricing

The practice of charging for road use in order to raise revenue and pay for infrastructure is ancient. Some toll roads and turnpikes go back centuries, and today governments levy taxes on vehicles and on petrol. But the idea of charging in order to deter road use and to prevent congestion is recent.

The aim is to ration road use by price. For instance, road use could be regulated by charging those who wished to drive on the busiest roads at the busiest times of day more than those driving at night on empty highways. One refinement to such models might be to let cleaner vehicles pay less, or exempt them totally, and make dirtier ones pay more.

Singapore started congestion charging in 1975 and such schemes are becoming more common in European cities, with two well-known examples in London and Stockholm: London's congestion charge is waived for electric vehicles. The technology involved in city-centre congestion charging is relatively simple, with cameras placed on the boundary of the congestion zone to photograph vehicle licence plates. A much more controversial idea is to extend this idea into national road-pricing schemes, which no country- apart from the small city-state of Singapore-has yet done.

National schemes would require satellite-based systems. Vehicles would have to contain a satellite tracking-device that would determine which roads were being driven along, for how far and at what time of day. This information would then be sent to a central computer system, and the appropriate charges levied against the driver. All these schemes, whether city congestion-charging or national road-pricing, meet stiff resistance. The Left often complains that congestion charging is regressive taxation hitting the poor hardest, while the Right frequently objects on libertarian grounds.

In the UK, the Labour government talked as recently as 2007 of introducing a national road-pricing scheme, as recommended by various expert reports. But the idea has stirred strong opposition from people who immediately placed a petition on the No. 10 Downing Street website – the petitions facility being an innovation introduced by the same

A cyclist in the central business district of Singapore cycles underneath the Electronic Road Pricing gantry.

Labour government – opposing road pricing. It garnered one million signatures less than three months after being posted.

Air travel

Biojet fuel

If many passenger cars can be converted to using electricity, then it would make sense for a low-carbon transport infrastructure to reserve biofuels for long-distance truck journeys beyond the range of even the best li-ion batteries. And for aircraft. While biofuels produce carbon emissions when burnt, this is carbon that the plants have taken out of the atmosphere when growing. Their use may not be carbon-free, but it is at least carbon-neutral. Moreover, no electric alternative exists for aircraft. (The US military carried out somewhat daft research, in the early days of the Cold War, into powering bombers with small atomic reactors, so that its nuclear- weapon-carrying bombers could stay aloft for long periods of times and not be caught napping on the ground in a surprise Soviet strike. The idea was dropped after it was realized that the same purpose could be achieved by putting nuclear weapons in submarines powered by nuclear reactors, and therefore capable of staying at sea for long periods.)

But biofuels are now being tried out for aviation. In the past two years there have been several commercial trials: Virgin Atlantic flew a Boeing 747 with one of its four engines running on a mix of coconut and babassunut oil; Air New Zealand used a fifty/fifty mix of jatropha oil and ordinary jet fuel in one of a 747's four engines; Japan Air Lines used a fifty percent mix of jatropha, camelina and algae in one of a 747's four engines, and Continental Airlines put a mix of fifty percent of jatropha and algae in one of a Boeing 737's two engines.

Apart from coconut oil – which Virgin openly admitted is not an ecologically desirable fuel-source due to deforestation issues – the other feedstocks look commercially and technically viable. Jatropha can be grown on very marginal land and camelina in temperate climes, while algae, which grows continuously, requires little land (see p.137). One absolute requirement for biojet fuel is that it does not freeze at high altitudes; some early biofuels had a higher freezing-point than conventional fuel, which was an obvious and alarming flaw, but one that is being overcome.

The aviation industry is very keen that any biojet fuel should be "feedstock agnostic", so that whatever source is used produces the same grade of biojet fuel. The industry is also insistent that biojet fuel should be capable of being used as a "drop-in" replacement for ordinary jet kerosene: in other words just dropped into the existing fossil-fuelled system without modifications to jet engines, or to the transport and storage of fuel.

Of course, the other way of reducing aviation emissions is to reduce the amount of fuel that planes burn. But this is proving increasingly hard to do as today's turbo-fan engines approach the limit of their potential. One technology, the so-called open-rotor engine, could save a lot of fuel, but is intrinsically noisier.

Emission trading

By deciding to include airlines in its Emission Trading Scheme, the European Union has devised a form of charging for the use of the airways over and around Europe. The aim of governments and of the EU is not to reduce air travel directly. Such a reduction would obviously be a plus for the climate, but it would run counter to aviation liberalization in North America and Europe, which has made flying cheaper (and, until 9/11, easier) on both sides of the Atlantic. The goal is rather to pressure airlines, aircraft makers and aero-engine manufacturers into reducing

An L-29 military aircraft powered solely by bio-diesel had a successful test-flight in 2007. Biodiesel Solutions and Green Flight International collaborated on the project, performed extensive testing on the ground with various bio-diesel and jet-fuel blended concoctions.

Finally, the pilot and crew deemed that the machine ran as normal using B100 fuel and was flown at a height of 17,000 feet above Reno, Nevada.

CO_2 and greenhouse gas emissions. Aviation emissions are currently only three percent of total greenhouse gases, but are the fastest-growing source. From 2012 the EU is to issue emission allowances for all flights inside, to and from, Europe. Initially, airlines will be issued enough emission allowances (85 percent of them free, the rest to be auctioned to the highest bidder) to cover 97 percent of the 2004–06 average emitted by aviation in and around Europe, with the amount of allowances decreasing in later years. Airlines will pass the cost of these emission allowances on to passengers.

Congestion – in the sky and on the ground

Congestion wastes fuel and causes unnecessary emissions. This is a big problem in crowded Europe. But both Europe and the UK could take plenty of steps to help themselves. Over the last decade, air traffic has grown by more than fifty percent. Europe now has close to 8.5m flights per year and up to 28,000 flights on busiest days. A large part of the problem is that there is no unified management of air traffic across Europe. This is something that the Eurocontrol air traffic control organization and the European Commission have been trying to solve with the Single European Sky programme.

The UK has a capital city burdened with the scatter-pattern legacy of having three airports with inadequate runways that they are unlikely, for political and legal reasons, to be able to increase in number. While most of their continental counterparts have four runways, Heathrow has only two, while Gatwick and Stansted have just one each. This causes congestion both in the sky and on the ground.

High-speed rail

It is not cut-and-dried that high-speed rail is the solution to saving energy and carbon emissions in the transport sector. The faster the train goes, the more energy it consumes, and the more CO_2 it emits per kilometre in order to overcome the greater wind resistance. Moreover, high-speed trains (meaning trains doing more than 200km an hour) usually require construction of a dedicated track with more gradual bends. The extra emissions resulting from the steel and cement in the construction of this dedicated track – compared to just upgrading existing conventional rail track – have to go into any proper calculation of the net energy/emissions balance of high-speed rail. But if this overall extra volume of energy and emissions involved in the construction and operation of high-speed rail results in a popular method of transport – if it can be divided by a large number of passengers – then the amount of energy or CO_2 attributed to each rail passenger falls to quite a low number, lower than for car or plane passengers.

The goal is to tempt enough drivers out of their cars (and passengers off their planes) and onto high-speed trains to make them enough of an energy-saver. It can be done. Japan paved the way with the Shinkansen bullet train between Tokyo and Osaka, followed by France with its *trains à grande vitesse* (TGVs) between Paris and Lyon. Both countries have a high proportion of zero-carbon electricity generation (from nuclear) to also make their high-speed trains a real emission-saver. In 2009 the UK's Labour government came out clearly in favour of "systematically replacing short-haul aviation with high-speed rail ... for reasons of carbon reduction and wider environmental benefit", in the words of Lord Adonis, the former transport secretary.

The French experience is that the threshold for "modal shift" (from one transport mode to another) that will persuade an air traveler onto a train is a TGV journey time of three hours. Research suggests that this threshold is moving, in rail's favour, towards four hours. One advantage of rail is that it is not subject to the same rigour of post-9/11 security checks that make air travel inconvenient. Another advantage is that rail-journey times are city-centre to city-centre. This is an advantage in Europe, which has generally good public-transport links inside cities linked to rail stations, though less often the case in the US, where people anticipate, rightly or wrongly, having to drive at their destination.

From an energy-saving perspective, the only foreseeable drawback of high-speed rail, were it to be successfully implemented using "green" electricity, could be that its convenience creates extra demand – the rebound effect we examined earlier.

Low-carbon energies
Picking winners

Renewable energy developers are rarely in it for just the money: usually the reward of knowing they are moving the planet in the right direction is the most significant factor. This is just as well. For low-carbon energy innovation is still not for the financially faint-hearted. Here are three related reasons.

Competition

This is intense. Low-carbon or renewable energy enters a sector dominated by fossil fuels. The infrastructure carrying fossil fuels – grids and pipelines – has been refined and optimized over the past century. So have the machines – notably the internal combustion engine – and they are engineered to use fossil fuels. The investment in both infrastructure and machines constitutes an enormous investment, that is hard to displace or integrate with.

Rewards

These are often low. For the biggest challenge of renewable energy is clean generation of electricity, which is a standardized commodity. Electrons are electrons and they will power whatever they will power irrespective of how they are generated. Few people will, of their own accord, pay extra to cover the higher early costs of a low-carbon technology that produces electrons which are indistinguishable from those produced by fossil fuels.

It is true some early "green" products, such as the Prius hybrid car, are chic enough to command a high price. But it is nothing compared with the readiness of people to pay hundreds of dollars for the distinction of owning an early mobile phone in the 1980s. Typically, to get their costs down, energy technologies need to take time refining their manufacturing and testing the marketplace, which allows imitators to get to work. There are few prizes for being first across the line. As if to underline this very point, Senator John McCain proposed, as part of his unsuccessful Republican bid for the White House in 2008, a $300m prize for the inventor of a really strong car battery, one capable of delivering the same or more power as current batteries at 30 percent of current costs.

Research

This is relatively feeble. Because the status quo in the rich nations is satisfactory in terms of delivering energy, leaving climate aside, and the rewards for altering it are slender, little public and private money has gone into energy research and development.

By the mid-1980s, once the scare of the 1979–80 oil shock had worn off, both governments and companies in the industrialized oil-consuming countries virtually ceased investing in energy R&D. Over thirty years, from 1974 to 2004, energy accounted for only six percent of US federal R&D spending. This is a trend President Obama is seeking to reverse. But most industrialized countries spend less than 0.03 percent of GDP on energy research; only Japan surpasses this, with a still meagre 0.08 percent.

Carbon penalties: the great equalizer?

The best way to level the playing field is a penalty on carbon: either a carbon tax, or a price rationing of carbon through tradeable carbon-emission permits. (Trading is a form of taxation, because the price carbon permits are traded at effectively constitutes a rate of tax.) The higher the carbon penalty, the quicker renewables can come on to the energy market. This is because new technology becomes cheaper through a series of what economists call learning curves: learning-by-searching, via R&D; learning-by-doing, via manufacturing; and learning-by-using, via consumer feedback. These curves (sloping downwards to the right on a typical graph) plot the way the price of a technology falls over time. The heavier the carbon penalty handicapping fossil fuels, the less the price of renewable technologies needs to fall to match fossil fuels in the marketplace.

Is this rigging the market in favour of renewables? Absolutely. "None of this [renewable technology] works except in a rigged market", says Dr Alex Buchan, technology director for North Star Equity Investors, a firm

A hydroelectricity-generating barrier in Näsåker, northern Sweden. Finland and Sweden are the only two nations in the world to impose a carbon tax (which the Swedish government increased by 2.6% in 2008 – to 2.34 kronor per litre). Sweden now generates all of its electricity by either hydropower or nuclear.

in Newcastle, UK, which has many new-energy technologies investments in its venture-capital portfolio. "My feeling is this market will remain rigged for a sensible length of time – there is consensus between the [UK] political parties on this. So the political risk is relatively limited – the issue is more the long lead times, especially with things like tide and wave power."

The "Valley of Death"

This is where money can dry up altogether and a project dies. If this happens, it is typically during the period *after* the government starts to phase its money out, but *before* the private investor can see enough light at the end of the tunnel to have confidence in getting a profitable return on his capital. This period can be long in renewable energy – making the valley of death seem alarmingly wide and giving the investor a financial panic attack.

That's partly why the Carbon Trust, a body set up by the UK government to act as a corporate midwife to low-carbon companies, cautions that the cost of sustaining and deploying low-carbon technologies and getting them through the valley of death can be up to forty times that of the initial R&D phase.

The recession made credit scarcer for renewable entrepreneurs. But another problem for renewable technologies is that they are very capital-intensive. Not necessarily more expensive than other projects. Indeed, because the fuel is often free wind, sun, tide, wave running costs are low. But this means that almost a hundred percent of the overall cost is in the capital equipment, which must be paid for upfront. New ways will have to be found of coping with this.

One way might be for financiers to put up the capital expenditure and be repaid by the project revenue. This is working with the most tried and tested of the renewable technologies – onshore wind. A turbine company called Wind Direct has formed a joint venture with HgCapital to erect turbines for industrial energy-users and to split the revenue with the energy user. HgCapital, the funders, are a large renewable-energy investor, with controlling interests in 35 European renewable energy projects. It's likely that we will be seeing similar large renewable investors emerging over the next decade.

The table overleaf, adapted from the IEA's *Energy Technology Perspectives*, demonstrates the scale and cost of the low-carbon challenge before us. The first column of figures gives an idea of how much CO_2 (in gigatonnes, or billions of tonnes) a range of technologies that already exist or are in advanced development would save in a scenario where the goal was to bring global emissions back to 2005 levels by 2050. This is not a very ambitious target.

On the money side, the assumption was that these technologies would not incur or be used in such a way as to incur top marginal costs of more than \$200 per tonne of CO_2 saved or avoided. So the total cost of the bill for using these 13 technologies (note that nothing is expected from hydrogen fuel cells in cars) would be between \$3.7 trillion and \$4.5 trillion. Hardly spare change, but not, in a global context, exorbitant.

But now look at the cost of doing what we think we know we have to do by 2050 – halve today's (or 2005) level of emissions by 2050. This would require utilizing, in the IEA's words, "technologies still under development,

ENERGY PROSPECTS

	CO$_2$ (gigatonnes)		US dollars (billions)	
	Return to 2005 level by 2050 (Scenario 1)	Cut to 50 per-cent below 2005 level by 2050 (Scenario 2)	Research, development, demonstration and deployment (RDD&D) (Scenario 1)	RDD&D costs (Scenario 2)
POWER	7.42	12.30	2,400–2,840	3,050–3,540
Carbon capture and storage of fossil fuels	2.89	4.85	700–800	1,300–1,500
Nuclear	2.00	2.80	600–750	650–750
On/offshore wind	1.30	2.14	600–700	600–700
Photo–voltaics	0.67	1.32	200–240	200–240
Concentrated solar power	0.56	1.19	300–350	300–350
BUILDINGS	6.98	8.24	320–400	340–420
Energy efficiency in buildings and appli-ances	6.50	7.00	n.a.	n.a.
Heat pumps	0.27	0.77	70–100	90–120
Solar space and water heating	0.21	0.47	250–300	250–300
TRANSPORT	8.20	12.52	260–310	7,600–9,220
Energy efficiency in transport	5.97	6.57	n.a.	n.a.
Second generation biofuel	1.77	2.16	90–110	100–120
Electric/plug–in vehicles	0.46	2.00	170–200	4,000–4,600
Hydrogen fuel-cell vehicles	0.00	1.79	n.a.	3,500–4,500
INDUSTRY	3.00	5.68	700–900	1,400–1,700
CCS for industry	2.00	4.28	700–900	1,400–1,700
Industrial motor systems	1.00	1.40	n.a.	n.a.
TOTAL	25.60	38.74	3,680–4,450	12,390–14,880

whose progress and ultimate success are hard to predict", such as hydro-gen fuel cells for cars (considered too undeveloped to be employed in the first scenario). Moreover, all technologies would be intensively deployed in such a way that the top marginal cost of saving or avoiding CO_2 might rise to $450 a tonne of CO_2. Using all 13 technologies this intensively to halve today's emission level would cost between $12.4 trillion and $14.9 trillion.

The table gives us an idea of where our two big priorities should be:

▶ **Energy efficiency** The table demonstrates the big efficiency savings that can be made in buildings. Likewise, big savings could be made by increasing the efficiency of conventional cars. However, notable also is the soaring cost of thoroughly decarbonizing the transport fleet via the widespread development and deployment of electric and hydrogen fuel-cell vehicles.

▶ **Power generation** The importance of making carbon-capture and storage work is clear, as is the significant, albeit more modest, contribution that wind, solar and biomass could make (the last is not featured in the table for reasons of space). Notice, however, how little extra benefit there would appear to be from a crash programme of nuclear power development.

What should the UK focus on?

In recent decades the UK governments, of whatever political complexion, have been allergic to the industrial strategy of picking winners. This is born partly out of the belief that if you set the right economic framework then a thousand flowers will bloom – the sort of approach which can work in a big, vibrant economy like the US, but which might be a little optimis-tically laissez-faire for Britain. And it is partly out of nervousness at the UK's past failure to pick winners – think of Britain's government-backed computer and car companies – as Germany has done with mechanical engineering and France with nuclear power. Nonetheless, the urgent necessity of acting to mitigate climate change, and the squeeze on public money, are causing a shift of attitude.

Dr Garry Staunton is Technology Director of the UK Carbon Trust, whose job it is to set priorities, has this to say on the matter. "It would be great if we had a large number of alternatives to pick winners from. But the flip side is that you have to drop losers, on which you may have spent quite a lot of money. I prefer to focus on reducing the number of runners." He has two in particular in mind.

Offshore wind

The UK has a total of eight Gigawatts (GW) planned or in operation. But this will have to increase to 29GW if the UK is to meet its renewable energy commitments to the EU. "In general wind-turbine development, the UK missed the boat, though we have had some success in blade-

---

A few good ideas

▶ **Solar cookers** Right at the start of this book, we underlined what a menace to the health of people and planet traditional biomass cooking can be, when precious trees are chopped down for cooking in badly ventilated kitchens. So it was particularly heartening to see a cheap-to-make ($5) solar cooker, designed by Kenyan John Bohmer and dubbed the Kyoto Box, win the $75,000 first prize in the *Financial Times* Climate Change Challenge, organized by the Forum for the Future sustainable development body.

Various models of solar cookers, functioning as heat traps and using the green-house effect to good purpose, have been around for some time. But the Kyoto Box has the great merit of simplicity and cheapness. It is composed of one cardboard box with aluminium foil to focus sunrays placed inside another cardboard

box, painted black to increase heat absorption, and topped with a clear acrylic cover to let in and trap the rays of the sun. Black paint, foil and insulation work together to raise the temperature high enough to boil water. By replacing firewood, the cooker could save up to two tonnes of carbon emissions per family per year. It could also save lives: it can boil ten litres of water in two hours, destroying the germs that kill many children each year in developing countries.

▶ **Road lighting** There is a law-and-order case for overhead street lighting especially in cities; criminals prefer to work in the dark. But street lighting consumes a lot of electricity – the only man-made features that astronauts are said to be able to spot on earth are China's Great Wall and Belgium's motorway network, lit up with off-peak night-time nuclear power.

design technology which has been bought by the Danes who have led the industry", Dr Staunton explains. "But the market in wind is beginning to split, into onshore, where US and Danish companies maintain their presence, and offshore, where the UK is a natural market because of its good wind-resources in relatively shallow coastal waters and its experience in oil and gas-rig work."

The Carbon Trust is particularly looking at ways in which the UK can become a leader in offshore wind-installation, operation and maintenance. This involves everything from modelling the optimum array of

And there is now another way of achieving the other purpose of street lighting – guiding drivers at night. A UK company called Clearview Traffic has developed "solar lite" road studs. These don't need to reflect headlights of cars but derive enough of their own energy, according to the company, from "a few hours of sunlight" to emit their own light from dusk to dawn for ten consecutive nights. These road studs are visible up to a kilometre and can light up the road some way ahead for the driver beyond the range of headlights. Whether or not these solar studs can or should entirely replace street lighting is a larger question, with the issue of crime on city streets also figuring in the debate. But there will be some roads where overhead lighting can be safely removed or reduced. Perhaps if or when street lamps are removed, a few electrical power points can be left behind for electric car drivers to recharge.

▶ **Computer efficiency** Computers can help us use energy more efficiently, but they can also waste it. What is particularly wasteful is the power that computers, and all sorts of consumer electronic gadgets, consume when they are not being used, but left in standby mode. There are a variety of EU proposals to require manufacturers to reduce their products' electricity appetite on standby. But a UK start-up company from Sheffield, with the cutsey name of Very-PC, already has remedies to offer. By assembling more efficient fans to keep computers cool and modifying the voltage of some components and the power management software, Very-PC says it can reduce the hundred watts per hour used by a computer that is turned on but doing nothing to less than thirty watts.

In an interesting aside, Very-PC's managing director, Peter Hopkins, was ridiculed in 2008 when he presented for his business idea for energy-efficient computers to the group of entrepreneurs on the panel of the *Dragons' Den* television programme. As the *Sunday Times* amusingly related, one of the "Dragon" entrepreneurs told Hopkins that he had "a pretty averagely crap business…this isn't Very-PC, it's very poor, you need to go back to the drawing board". Luckily, Hopkins did not take this advice, and is now running a thriving business. That is an uplifting lesson for any down-hearted green-energy developer.

turbines at sea and researching alternative ways of mooring and anchoring turbines, to finding easier, safer and cheaper ways of getting engineers on and off sea turbines. "So while the UK may not become a world exporter of the main technology, it could develop expertise in the supply chain, and innovate ways of making operation and maintenance leaner, which could be sold to the world." Marine energy – meaning wave and tidal power – is a less pressing matter, though Dr Staunton says "it has the same potential as wind was spotted as having twenty years ago".

Next-generation solar PV

The Carbon Trust is also very interested in solar voltaics, which might initially seem surprising. "Silicon has been the base of this industry. It has been very expensive", Staunton explains. "A lot of government subsidy has gone into supporting it, but outside the subsidized markets of mainly Germany, Spain and the US it is not very attractive." In contrast to wind, the UK is hardly a natural market for solar power.

"But if you look ahead to the next stage of PV generation, where the search is on for materials such as polymeric, plastic materials that can be printed off cheaper than silicon, then the UK has a chance", Staunton says. "Because we have people who know about chemicals, printing and plastic electronics."

Your flexible friend: this strip of "power plastic", manufactured by the Massachusetts-based firm Konarka, is an example of so-called third-generation solar photovoltaic technology. A rival firm, Solarmer Energy, currently holds the record for conversion efficiency in printed plastic solar cells – at 7.9 percent.

Environmental politics

Resistance and pressure for change

The pressure for a new-energy economy has largely come from the environmental movement. Environmental organizations are now active around the world, with groups conducting very important campaigns even in countries such as China. In this section, however, we concentrate on environmental campaigning in Europe and the US, because it is here through climate policy that environmental campaigning has had the most transparent and direct effect on the energy sector.

Britain, for instance, has passed legislation in 2008 committing all future UK governments to working within five-year carbon budgets so as to reach legally binding carbon-emission reduction targets. This was a direct result of environmental NGO pressure (see box on p.252). In this chapter, we take a look at how the major energy sources of today – and their counterparts in renewables – are regarded by environmental groups.

Coal: the new nuclear

Increasingly, environmental protests have targeted coal plants across the world – in Australia, Germany, the UK and many states in the US – because of coal's heavy carbon emissions. In the US, the Sierra Club started its Beyond Coal campaign at the start of the twenty-first century, pledging to "leave no new coal plant unopposed". It vaunts considerable success in this, claiming that, of the 150 coal power plants proposed since 2001, no fewer than 111 such projects have been "defeated or abandoned", and that in 2009, for the first time in six years, no work was started in the US on any new coal project.

A brief history of environmentalism

The world's first environmental group could be said to be the Commons, Footpaths and Open Spaces Preservation Society, founded in Britain in 1865. The Sierra Club, the most influential US environmental group, was established in 1892. But the environmental movement has many disparate offshoots.

Probably the most radical approach to easing the strain on the earth's environmental and energy resources cannot be called environmental at all – it is the Voluntary Human Extinction Movement (pronounced "vehement", from its website vhemt.org) which believes the best thing that human beings can do for the planet is to stop breeding. Environmentalism is certainly not monolithic, and has many strands to it – anti-capitalism, anti-globalization, eco-feminism (which blames the rape of the environment on masculine vices of greed and aggression) and conservation (represented by organizations such as the Sierra Club in the US or the Campaign to Protect Rural England and the National Trust in the UK).

Different environmentalists adopt different tactics. Some lobby governments, while others stage spectacular protests. Almost all environmental groups are non-violent, though a few – Earth First and Earth Liberation Front indulge in deliberate sabotage (or "eco-tage") of tree-cutting or road-making equipment. If the worst prognoses about climate change are correct, we will probably witness an increasing radicalization of the green movement: eco-terrorism could well become a regular feature of the headlines of the future. Others carry out direct action with a sense of humour, such as the members of Greenpeace in Australia who put solar panels on the house of John Howard, the country's climate-sceptical former prime minister.

For all the complaints of some environmentalists about globalization, environmentalism as we know it today is undoubtedly a global movement that has its roots in North America. Many environmental organizations started there and spread their offshoots to Europe. Greenpeace began as a Canadian-based organization protesting against US nuclear testing off Alaska. It now has branches in 46 countries affiliated to Greenpeace International, based in Amsterdam. Friends of the Earth started as a splinter from the Sierra Club in the US, but now has branches in 77 countries. Earth First began in the US, and was established later in the UK.

The focus naturally differed on either side of the Atlantic. Tree-sitting started in California as a means of stopping the felling of giant redwoods, in protest at which Julia "Butterfly" Hill spent a record 738 days up a tree. Tree-sitting, as well as tunnelling, was a tactic used in the UK, only there it was in order to attempt to prevent the practice of putting town by passes through local beauty spots, a smaller-scale despoliation of landscape that mattered less in the bigger US.

However, some US green groups remain uncertain about what stand to take on newer technologies (carbon capture and storage, or CCS) designed to clean up coal-fired power generation. The Environmental

Defense Fund has been ambivalent on CCS, while the Natural Resources Defense Council still claims that the coal industry's advertisements for "clean coal" technology are "a fraud" and that any use of coal remains "dirty and dangerous" for the climate.

In the UK, environmental protesters have frequently targeted Drax, the country's largest coal plant – and carbon emitter – while protesters have also attempted to detain trains and ships taking coal to power stations. In 2008–09 one particular UK target for protestors has been Kingsnorth in Kent, where Eon, a leading energy company, planned to add a large new coal-power plant to an existing one. In 2007 protestors entered the existing coal plant, climbed a 200-metre chimney and painted the word "Gordon" [for Gordon Brown, the then prime minister] on it. (After their epic climb, they were unable to finish painting the full message they had intended.) A year later, six of the protestors were cleared of criminal-damage charges in a hearing in which the defence called US climate-change scientist Jim Hansen and an Inuit from Greenland to testify as witnesses.

It was believed to be the first case in which the defence acknowledged the damage, but successfully argued it was justified on the grounds of try-

Greenpeace activists protest against carbon emissions in front of a coal-fired power station west of Beijing in July 2009. Greenpeace claims that greenhouse gas emissions from China's top three power firms outrank those of the entire United Kingdom.

ing to prevent greater damage to the climate caused by Kingsnorth belching carbon. Part of their defence rested upon complicated calculations of the amount of emissions saved by their occupation. The Kingsnorth case brought the issue of cleaning up coal with CCS to public attention. But before it could become clear whether Eon would volunteer to equip Kingsnorth with CCS, or whether the UK government would require it, Eon decided to drop the expansion of Kingsnorth, saying that it didn't see sufficient market demand for the extra electricity.

In general, the cause of climate-change control – and at the heart of it, pressure for low-carbon energy – has mobilized a very wide range of NGOs to work together in the UK, as elsewhere in the world. Most of the NGOs in the UK fall into three main categories:

▶ **Green groups** with a broad concern about the environment and biodiversity. These include animal welfare groups such as the Royal Society for the Protection of Birds (RSPB) and WWF (formerly the World

Police surround protesters during a sitdown protest at the gates of Kingsnorth Power Station near Rochester in Kent, in August 2008.

Wildlife Fund).

▶ **Development organisations** such as Oxfam, Christian Aid and Tearfund. Their prime concern is that climate change will hit the poorest developing countries hardest.

▶ **Landscape-preservation organizations** such as the National Trust and the Campaign to Protect Rural England (and their Scottish and Welsh counterparts). Climate change poses clear risks, such as flooding and coastal erosion, to the UK countryside.

These groups complement each other in membership, finance and tactics. The RSPB brings a one-million-strong membership (more than any other NGO or indeed any UK political party) to the cause, although it has to be said that most of its members joined to visit their local bird sanctuary rather than to campaign on climate change. The inclusion of the overseas-development organizations often taps into a religious constituency, in the case of Christian Aid and Tearfund (founded by evangelical Christians). The development organizations also have bigger budgets, although, as far as climate change goes, these are focussed more on helping the world's poor adapt to climate change than on campaigning to prevent it.

A coalition of NGOs brings a range of tactics to the fight – and that can be useful. For there is a "good cop, bad cop" synergy between quintessentially "outside" organizations such as Greenpeace, which specialize in direct action and shock or surprise tactics, and environmental groups which function on the "inside" of political establishments. One example of the latter in the UK is the Green Alliance. It is an environmental think-tank which works very closely with UK politicians and civil servants in trying to turn green proposals, often originating with other green NGOs, into policy and law. But its director, Stephen Hale, is clear that Green Alliance owes much of its success in Westminster and Whitehall to pressure from outside groups. "Inside-track organisations like ours will ultimately only succeed if there is pressure for action building from the outside", he says.

Nevertheless, there are certain energy sources that are harder for environmental groups to reach a consensus on.

Wind

The "landscape lobby" – particularly the National Trust and the Campaign to Protect Rural England – remain strongly opposed to most onshore wind farms. Because of the large amount of land it owns or controls around the UK, the National Trust can be a very powerful opponent on onshore renewable energy; it usually only backs micro-wind, offshore wind and "run of the river" hydropower projects that don't involve dams.

On the other side of the fence, Greenpeace and Friends of the Earth are strongly in favour of wind power, whether on-or offshore. Objections to wind farms are often localized. It is therefore intrinsically easier for Greenpeace to take a pro-wind stance, because it is an organization that has a national membership but no local campaign groups. Friends of the Earth UK favours wind, despite having some 250 local groups. The reason, says FoE climate-change coordinator Mike Childs, is that "our watchword has always been 'think global, act local', so we have first thought about what is needed globally – more wind power – and then applied it locally". He adds that FoE is "not part of the landscape movement".

A big green success: putting climate on the statute book

The 2008 Climate Change Act is a far-reaching piece of legislation. It establishes legally binding targets for the UK to cut greenhouse gases (from the 1990 level) by 80 percent by 2050 and carbon dioxide by 26 percent by 2020.

It also sets out a road map of how to get to these goals, via five-year carbon budgets and the establishment of a climate-change committee to advise how these budgets should be set. This piece of legislation is very much the result of pressure from environmental NGOs, and of one in particular – Friends of the Earth.

▶ **2005** FoE drafted a climate-change bill, calling for annual cuts in carbon. In 2005, FoE announced that it had grown impatient with the Labour government's failure to act on its successive election-manifesto pledges on climate change, and disappointed by the failure of government departments such as transport to do anything about their big carbon footprint. It designed the draft bill to have year-on-year carbon reductions so as to avoid the "tendency of politicians to hope for some technological breakthrough that would bail them out". FoE launched the Big Ask campaign – to request the government to take adequate climate-change action – and persuaded Radiohead singer Thom Yorke to head it. "Getting celebrity support opens you up to other audiences, makes your campaign interesting to the media – and Radiohead had a reputation as the 'thinking man's band'", says Mike Childs, FoE's climate-campaign coordinator. At the same time, the Stop Climate Chaos coalition was formed to give wider backing to the climate bill. It included many other organizations such as WWF, Greenpeace, RSPB, Oxfam, CAFOD and the Women's Institute.

A striking example of a national organization overriding local members' views is the RSPB. Not surprisingly, the RSPB has traditionally been against wind turbines because of the birds that accidentally fly into them, and most RSPB local groups still oppose them. But the RSPB national organization commissioned a report, published in March 2009, which called for more wind farms in the UK, provided these were not in areas identified on "bird sensitivity" maps as being of special concern to bird conservationists. With this safeguard in mind, the report said "we need a clear lead from government on where wind farms should be built".

Nuclear

Nuclear power is still controversial among environmental groups, although some of them, such as WWF or the RSPB, prefer to stay silent on the sub-

▶ **2006** Thom Yorke headlined the Big Ask Live concert in London, attended by David Miliband (who was about to become Labour government's environment minister) and David Cameron (the leader of the Conservative opposition). The support of both these men was thought to be crucial. Cameron, who was re-branding the Tories as a greener, more caring party, came out in support of a climate bill that was almost identical to that proposed by FoE. Even more important was the subsequent replacement of Margaret Beckett as Environment Secretary by David Miliband. "Whereas Beckett saw it as her job to keep NGO pressure off her department and the government, Miliband's approach was to say 'you bring the pressure on and I'll get the bill through the cabinet', according to Mike Childs. No fewer than 412 MPs (out of a total of 646) signed an early-day motion supporting a climate-change bill. FoE had turned up the pressure on MPs via a grassroots campaign, sending MPs postcards urging them to vote for the motion.

▶ **2007** A draft Climate Change bill was unveiled by the government. But FoE and other NGOs kept up the pressure to get the bill strengthened. It was. The emission-reduction goal was increased to an eighty percent cut by 2050, and emissions from aviation and shipping were included.

▶ **2008** The Climate Change Act became law. How workable the legislation will be is another matter. What do legally binding targets really mean? Can a government bind its successors? How would a government discipline itself for falling short of a target? One thing is certain – the bill's clauses rendering the targets as "legally binding" provide the opportunity to take governments to court – an opportunity that environmental groups will doubtless take.

ject these days. Very few environmentalists actively favour nuclear power. James Lovelock, who put forward the "Gaia hypothesis", is an important exception. He is now an outright advocate of a large-scale programme for nuclear power, because of the sheer amount of low-carbon energy it can contribute in the fight against climate change. Other environmentalists accept that nuclear power's contribution to low-carbon energy is beginning to neutralize nuclear power's economic and safety handicaps. George Monbiot, perhaps the most prominent environmental polemicist writing today, falls into this category of reluctant converts to nuclear power, provided that it can be proven that the resulting waste can be safety disposed of.

One organization, Greenpeace, remains viscerally opposed to nuclear power. This is not in the least surprising, given the organization's origins in campaigning against nuclear weapon tests and the sinking of its ship *Rainbow Warrior*.

(In 1985, French secret service agents sank the *Rainbow Warrior* in the harbour of Auckland, New Zealand, resulting in the death by drowning of a photographer, in order to prevent the vessel sailing on to obstruct a French nuclear-weapons test in the south Pacific).

However, not all of those who work or have worked at Greenpeace believe that outright opposition to nuclear power is an article of faith. Stephen Tindale, a former Labour governmental adviser, was head of Greenpeace UK for six years. But after resigning from that post, he came out in favour of

The damaged and no-longer functional Unit A at the nuclear power plant of Gundremmingen. Unit A was the site of the first fatal accident in a nuclear power plant and, subsequently, of a major accident resulting in the unit's total loss, the only to date in a nuclear power plant in Germany.

nuclear power in 2009 as the lesser of two evils. "My main reason for this was my conclusion that surface transport has to go electric", he explained. "So we need more electricity. But where is this coming to come from? Which is worse – unabated coal [without any CSS] or nuclear power?" He disagrees that nuclear is a distraction from, or the main rival to, the renewables industy. "The fact that France with all its nuclear power does better than the UK on renewables – not just on biofuels but better on wind than the UK – disposes of that argument."

But most environmentalists continue to quietly oppose nuclear power on practical grounds. "We're not pro-nuclear," says Stephen Hale of Green Alliance, "not because of ideology, but because it's just not cost-effective." Mike Childs of FoE adopts the same tone of low-key opposition. "We're not actively campaigning against nuclear, and we're not ideologically opposed to it or to any particular technology. But the timescales are now too short [in the UK] for nuclear to make a difference: we need to draw on technology that is faster to put in place."

Peak oil

With oil being an unredeemably GHG-generating fuel, its most pressing concern – peak oil – is not an issue that divides environmentalists as wind and nuclear power do. Indeed, all environmental groups welcome any signs of a peak in oil supply, and they positively campaign for a peak in oil demand (in that they try to persuade people to use ever less oil). But there is a division in the environmental movement on how much publicity to give to peak oil supply.

There are those who believe that peak oil is a more immediate and powerful lever than climate change in getting people to mend their fossil-fuel-dependent ways, and that therefore peak oil should be given headline treatment. There has been a minor flood of books written on both sides of the Atlantic about imminent peak oil. The best known peak-oil environmentalist is Rob Hopkins, who has tried to turn his home town of Totnes in Devon, England into a model for the "transition town" movement – an idea that has caught on in many other towns across the UK. The idea is to attempt to move communities towards a "re-localized" economy more dependent on local produce and services and therefore more resilient in the face of the coming peak-oil shock.

"One of the things peak oil does very effectively is put a mirror up to a community and ask: 'What has happened to the ability of this community to provide for its basic needs?'," writes Hopkins in his *Transition*

The mountains in Rifle, Colorado, are rich in reserves of oil shale. Environmentalists consider exploitation of "difficult oil" to be beyond the pale: both for its perpetuation of fossil fuel reliance and for the surface despoliation its mining will cause.

Handbook. "Allowing people to mentally explore what their current life-styles would be like if the inflow of cheap oil were to cease is a powerful way to get people to think about the vulnerability of their oil-dependent state. It can focus the mind more than climate change because it can seem to be more obviously relevant to people's everyday lives." One of the things that Hopkins appears to find appealing is that the peak-oil campaign has no government endorsement – in contrast to climate-change campaigns – and therefore the transition movement can be a bottom-up movement, not top-down.

Then there are those who fear that publicity about peak oil might panic the world into replacing exhausted stores of conventional oil with dirtier unconventional oil from Venezuela and Canada, rather than mobilizing renewable energy. Mike Childs of Friends of the Earth explains why many green organizations, if anything, underplay peak oil. "We do not focus much on the issue, partly because of the possible ambiguous response to peak oil", he says. "The other reason is that if you *do* focus on peak oil, you enter into a debate about geology in which there is no solid level of evidence as the Intergovernmental Panel on Climate Change has built up on climate change".

Jeremy Leggett sees it as one of his roles to bridge the gap between the peakists and the greens. As chief executive of Solar Century, he makes a good bridge: he was an oil geologist who became a Greenpeace policy activist before starting his own solar-development company. "There is a curious cultural relationship between the peak-oil folk and environmentalists", Leggett notes. "Some in the peak-oil camp are very dismissive about environmentalists and just uninterested in climate change. While some environmentalists are very resistant to the arguments of the peak-oil geologists. This is partly a cultural resistance to having the stature of your bogeyman [the oil companies] diminished by arguments you don't understand. But it is also partly a fear that the argument about conventional oil peaking will accelerate the panic into [using] unconventional fossil fuels like Canadian tar sands."

The green movement's nervousness about tar sands is understandable. The 2009 recession has eased immediate concerns about full-throttle development of tar sands. The International Energy Agency estimates that some 1.7mb/d of the deferred capacity has been in Canadian tar sands, out of a total capacity of 2mb/d of recession-deferred projects. Notwithstanding this deferral, the green movement in general has a barely veiled contempt for the desperation shown by oil companies in pursuing oil-or tar-sands development. As Rob Hopkins of the Transition Town movement colourfully puts it, extracting oil from tar sands is like trying to suck old beer out of the pub carpet after the bar has run dry.

Partly to get peakists and greens to think more about each other's preoccupations and the consequences of these preoccupations, Leggett helped to form the UK Industry Taskforce on Peak Oil and Energy Security, which includes representation from the engineering (Arup), utilities (Scottish and Southern Energy) and transport (Stagecoach and Virgin groups) sectors. It produced a report in autumn 2008 called *The Oil Crunch*, calling on the UK government to face up to the peak-oil problem, to expand oil and gas production in the short term, to accelerate use of biofuels and electricity in transport, and to move ahead with nuclear and renewables.

Expansion of oil and gas output – even just for the short term – looks like heresy to greens. But, for Leggett, eking out conventional oil and gas a bit longer is, along with investing in low-carbon alternatives, better than an unplanned plunge into polluting tar sands or unfiltered coal. In other words, it's better to plan than to panic.

The US oil industry and climate legislation

A guerilla war?

America is addicted to oil, according to former President George W. Bush, and kicking the habit is going to be harder there than anywhere else. In the summer of 2009 the US oil industry mobilized to oppose climate legislation in Congress, just as it had done when the Kyoto Protocol was being negotiated in the mid-1990s. As the administration-supported Waxman-Markey Climate Change Bill was passing through the House of Representatives – which it did by margin of just one vote – the president of the American Petroleum Institute, representing the US oil and gas industry, laid out his plans to up the pressure on Senators to kill the legislation.

The Waxman-Markey bill

Given its track record on previous climate legislation, it was hardly surprising that the API should oppose and actively campaign against the Waxman-Markey Bill. What was surprising was the API's suggestion to oil-company bosses that they should more or less direct their employees to take part in anti-climate demonstrations. In a confidential memo to API CEOs obtained by Greenpeace, Jack Gerard, the API president described the "Energy Citizen" rallies that had been planned around the country during Congress's August recess.

"The objective of these rallies is to put a human face on the impacts of unsound energy policy and to aim a loud message at … Senators to avoid the mistakes embodied in the House climate bill and the Obama administration's tax increases on our industry", wrote Gerard. "Please indicate to your company leadership your strong support for employee participation in the rallies", adding that their "facility manager's commitment to provide significant attendance is essential to achieving the participation level that Senators cannot ignore". This amounted to what the API's critics called "astroturfing": the creation of a bogus grassroots-campaign to influence Congress on energy and climate change. (A similar campaign was being waged at the same time in the campaign against Obama's proposed healthcare reforms.)

The Gerard memo re-opened the transatlantic rift between oil companies over climate change that first became evident in the mid-1990s. At that time most of the oil majors belonged to an anti-climate action lobby

The two authors of the US bill to establish a cap-and-trade plan for greenhouse gases: Henry A. Waxman of California and Edward J. Markey of Massachusetts, both Democrats.

group called the Global Climate Coalition. This group's main function was to stress and fund research that stressed the scientific uncertainty about climate change.

But in 1996 BP, a UK oil major with a big North American presence, quit this coalition, saying it could no longer deny the gathering evidence about global warming and about the role of fossil fuels in climate change. This was the first pro-climate stand by any oil major, and it led to defections from the Global Climate Coalition by other Europe-based oil majors such as Shell and, eventually, the dissolution of the Global Climate Coalition in 2002.

In 2007 the US Climate Action Partnership was formed among a variety of companies. Some of them were in the energy field: utilities such as Duke Energy, NRG Energy and Exelon, and oil companies like BP America, Shell and ConocoPhillips. As its name suggests, USCAP favours action. In August 2009 – just as the API's Jack Gerard was sending out his memo – USCAP had publicly reaffirmed the need for cleaner energy as "a necessary investment in the future". The Gerard memo being made public came as an embarrassment to the two European oil-company members, BP and Shell, which belong to USCAP as well as to the API. Subsequently BP and ConocoPhillips quit USCAP, leaving only one oil major, Shell, as a member.

Egregious Exxon

For years the most consistent energy-sector opponent of climate action has been ExxonMobil. Indeed under Lee Raymond, its CEO between 1993 and 2005, Exxon almost seemed to delight in its villainous reputation among the environmental community. It has been accused by America's Union of Concerned Scientists of circulating deliberate disinformation on climate change. "In an effort to deceive the public about the reality of global warming, ExxonMobil has underwritten the most sophisticated and most successful disinformation campaign since the tobacco industry misled the public about the scientific evidence linking smoking to lung cancer and heart disease", said the scientists in their 2007 report. By "underwritten", the scientists said they were referring to the fact that "ExxonMobil has funnelled about $16m between 1998 and 2005 to a network of ideological and advocacy organisations that manufacture uncertainty on the subject".

Raymond's successor as Exxon CEO, Rex Tillerson, appeared to soften the company's climate stance. In 2006 the Royal Society, Britain's oldest and most prestigious scientific association, wrote to ExxonMobil asking it to stop funding bodies that were putting out misleading information on climate change. The Royal Society claimed that ExxonMobil had promised to stop financing bodies that diverted attention from the need for cleaner energy.

While ExxonMobil has always been the environmentalists' villain, its solidly reliable financial and technical performance has made it the darling of oil companies for investors. In financial terms it has consistently outperformed its closest rivals, BP and Shell. One reason might be that, alone among oil majors, Exxon has not until very recently invested in any renewable energy projects (which, as explained in chapter

Lee Raymond, former CEO of ExxonMobil, argued that the company, with its technical and financial strength, could afford to let others take the lead on renewable energy but still catch them up if or when renewables proved commercial.

10, are expensive to set up). In 2009 Exxon took its first step in this direction, teaming up with gene scientist Craig Ventner to invest in biofuels from algae.

The public, the politicos and the celebs
Public figures go green

Celebrities set trends. But while that's true of clothes and harstyles, when it comes to energy use and climate change, public figures – be they film stars as well as politicians and businessmen – tend to be fashion followers rather than fashion leaders. Going green now forms part of many public figures' PR strategy. But is this so-called "greenwash" such a bad thing?

Politicians

Arnold Schwarzenegger is probably America's most colourful convert to green politics, having had a big impact in both image and substance. His change of tack is famously wrapped up with the fate of the Hummer, America's bulkiest sports and utility vehicle (SUV), which seems almost designed to be as boxy, and therefore as unaerodynamic and fuel-inefficient as possible. In 1992, the year after Schwarzenegger's biggest box-office success with *Terminator II: Judgement Day*, the AM General car company brought out a civilian version of the HumVee all-terrain vehicle it was making for the US military. Schwarzenegger was virtually the first customer, and he went on to buy several versions of the vehicle, which seemed to suit his image to a tee.

His patronage helped the Hummer become a commercial success, despite its inability to manage more than ten miles to the gallon in the city, and in 1998 AM General sold the brand to General Motors. Eventually sales were hit by the 2008–09 recession, and in June 2009 GM, as part of its bankruptcy deal, announced it would discontinue the Hummer. One final irony was that a Chinese company agreed to buy the Hummer brand and manufacturing equipment, but Beijing blocked the deal on environmental grounds.

By this time, however, Schwarzenegger had had a Pauline conversion on the road to Sacramento (the state capital of California) and become one of the greenest governors that California has ever had. He con-

The Hummer: a potent symbol of gas-guzzling fuel inefficency. Governor Arnie's former favourite military vehicle is no longer manufactured in the US.

verted one of his Hummers to run on hydrogen, and another to biofuels. Under Schwarzenegger, California has resumed its pioneering US role in progress towards a low-carbon economy. He signed legislation in 2006 setting a mandatory cap on greenhouse-gas emissions in California, the first such move in the US.

Al Gore's tendency to take himself a mite too seriously was considered a weakness as a politician, and he has been dubbed the "Goracle" by some, both affectionately and pejoratively. But his seriousness of purpose appears eminently suited to campaigning for climate action. It is not so much that he was a particularly successful Vice President (to Bill Clinton) from 1992 to 2000: the Clinton administration did not have an especially distinguished environmental record. It is more that very many people considered that, as the Democratic presidential candidate in 2000, he was the moral victor of that election, because he won more votes than George W. Bush even though the latter triumphed on the arcane arithmetic of the electoral college.

Gore's moral authority to speak out on the environment was further enhanced, in the eyes of many, by the Bush administration's rejection of the evidence of climate change, of the Kyoto Protocol and of the need to take any substantive climate action. The only drawback of having such a

Hollywood goes green – setting fashion or following it?

It is easy to ridicule many of today's big-name celebrities for empty gestures or double standards. A classic case was the report that a Lexus hybrid car was delivered to Paul McCartney by private jet from Japan. Yet there is a bigger and positive point to be made about "greenwash". The fact that public figures wish to vaunt their "greenness" shows how greenness has become not just socially acceptable, but a badge of cool. Moreover, celebrity involvement certainly attracts media attention, which in turn helps to spread the message and steer some of us in the right direction.

Some of the Hollywood elite have been making green lifestyle-choices or supporting green causes for most of their lives. They include Robert Redford (who has worked for conservation causes from his Utah mountain fastness), Ed Begley (who turns up at Hollywood events on his bicycle), and Edward Norton (who teamed up with BP Solar to create the BP Solar Neighbors programme, which aims to provide solar power to poor families in Los Angeles).

Of course, what film stars can do for the low-carbon cause through the power of film is even greater than the power of personal example. Leonardo DiCaprio has narrated a number of environmental films, *Global Warning*, *Water Planet* and most recently *Eleventh Hour.*, and has his own eco-activist website.

A recent British film phenomenon has been *The Age of Stupid*, a film about the impending catastrophe of climate change. Director Franny Armstrong raised the money to make it from the general public, an exercise in "crowd-funding". She approached Pete Postlethwaite to star in it, after an Internet search revealed that he was trying to get permission to put a wind turbine on his roof.

By making the film a success, Postlethwaite will have done much more to combat climate change than installing a mini-turbine (which, as we saw in Chapter Six, is rather inefficient). Capitalizing on her film's success, Armstrong went on, in August 2009, to launch the 10:10 campaign, to encourage people to sign up to cut their own emissions by ten percent by the end of 2010.

She received personal pledges more or less immediately from leaders of all three main UK political parties – the entire Labour government cabinet, the shadow Conservative cabinet and all the Liberal Democrat frontbenchers. A cynic would point out that 2010 was an election year in the UK. Nonetheless, Armstrong described herself as astonished. "It's amazing that within 48 hours of the campaign's launch, the leaderships of the three main political parties have committed to cut their 10 percent. Who said people power was dead?"

prominent Democrat play such a large part in promoting climate action is that it may have further solidified Republican opposition to climate action. But there is no doubt that he owes his successful climate-change campaign, his film (and book) *An Inconvenient Truth*, and the 2007 Nobel

Peace Prize that he shared with the Intergovernmental Panel on Climate Change, to having been a prominent US politician.

On the other side of the Atlantic, energy and the environment have risen to the top of the UK political agenda. But, since Britain's major political parties are not too far apart on climate policies, the contest has to some extent boiled down to the "personal greenness" of party leaders. David Cameron has made the environment a major aspect of his re-branding of the Conservative party, not only in terms of policy but also in personal image.

Shortly after becoming Conservative leader, he was filmed and pho-tographed riding a dogsled on a glacier in the Norwegian island of Spitsbergen to draw attention to the problem of climate change where it was beginning to have an effect. He had a wind turbine installed on his London house (which he had to take down briefly because it infringed local planning rules). But he was only really tripped up by the discovery that his well-publicized bicycle commutes to the House of Commons were facilitated by a chauffeur bringing his briefcase and extra clothes along behind in a car. Gestures of this kind are not the strong suit of the former prime minister Gordon Brown – who, however, did let it be known that he had "quietly" put solar panels on his house in Scotland.

Business

A wide range of companies have been re-branding themselves as green, and many of the world's wealthiest business people have been investing in clean energy. So much so that when the *Sunday Times* published its first ever "Green Rich List", in 2009, it did not look very different from the ordinary "Rich List" it has published for the past twenty years. The list was headed by Warren Buffet, one of the world's richest men. The sheer scale of his investments placed him at the top: they include some US utilities develop-ing wind power and a Hong Kong company making electric-car batteries.

The green business celeb who has undoubtedly attracted the most atten-tion – and criticism – is Sir Richard Branson. He has been industrious in his efforts to prevent the GHG emissions of his fleet tarnishing the image of his airlines (and the rest of the Virgin Group). In 2006 he pledged that all of the Virgin transport division's profits would be ploughed back into alternatives to standard kerosene jet-fuel. The degree to which this has actually happened is hard to gauge because Virgin is a private group.

But Branson has experimented with running Virgin planes on biofuels and is interested in investing in algae biofuels. At the same time, however,

Branson has exposed himself to charges of hypocrisy by backing a third runway at London's Heathrow airport and of creating Virgin Galactic to promote space travel.

Energy conspiracies

The energy sector has historically been vulnerable to conspiratorial practices. The 150-year history of oil has been marked by collusion, ranging from the Rockefellers' Standard Oil Trust, which was broken up under US anti-trust law a century ago, to today's international cartel of OPEC oil producers. But it also lends itself to an extraordinary number of conspiracy theories. Perhaps it is because we have a need for energy that is almost as universal as our need for air: we'd like it to be free as air too, and some of us feel that there must be something wrong because it isn't.

We have already looked at the part that lobbies, cartels and alliances play in today's climate-change debate, particularly in the US. The fossil fuel lobby has called global warming a hoax concocted by greens and scientists to justify closing down the oil, gas and coal industries down – and for their part, climate activists have labelled their opponents climate-change deniers, a phrase intended to smack of holocaust denial. But energy conspiracy theories pre-date climate change: they go back a long way.

Typically they involve alliances of governments and special interests alleged to be conspiring to suppress some innovation in "free energy" in order to maintain a status quo based on increasing fuel prices. And if the inventor or promoter of the "free energy" dies suddenly – as has happened in a couple of the cases cited below – suspicions of foul play fuel the conspiracy theorizing.

▶ **Free energy** We owe numerous electrical inventions, including the alternating current that powers most electrical motors, to the brilliant Nikola Tesla, who pioneered electricity in the US in the late nineteenth century, along with Thomas Edison. But in later life Tesla, who only died in 1943, theorized about many other things – among them free-energy devices and power sources able to propel aircraft and flying saucers. A few of Tesla's disciples, especially American physicist and author William Lyne, believe that the US authorities hushed up Tesla's power-source theories for fear they would disrupt America's oil and auto industries.

Some people believe that the US government may still be in the business of "free energy" suppression. Gary McKinnon, a Scottish-born computer hacker, is facing extradition from the UK to the US on charges of perpetrating what one US prosecutor has called "the biggest military computer hack of all time". He is said to have hacked into no fewer than 97 US military and NASA computers. Asked about his motivation, McKinnon cited an anti-government initiative in the US, called the Disclosure Project, which alleges a US government cover-up of information about Unidentified Flying Objects (an echo of Tesla?) and about "free energy".

▶ **Electricity storage** The US government was nearly taken in by Armenian-born "inventor" Garabed Giragossian, who claimed that, using an area no bigger than Boston Common (48 acres), he could develop enough free power to drive all the world's industrial machinery. He sought legal protection for his "invention" from Congress and was duly granted this in a resolution signed by President Woodrow Wilson – providing that he could prove the practicality of his discovery. It turned out that he had nothing more in mind than a giant flywheel releasing a spurt of energy as it unwound.

However, flywheel rotors have been developed as a source and store of energy, and are now being introduced into, among other things, Formula 1 racing cars. In order to be seen to do something for the environment, the Grand Prix racing industry has allowed Kinetic Energy Recovery Systems to be fitted to racing cars. These gadgets recover the energy lost in braking – either mechanically, with flywheels, or electrically with super-batteries – and re-use it as a boost to acceleration. In July 2009 Lewis Hamilton became the first driver to win a Grand Prix with a car fitted with KERS.

▶ **Killers of the electric car** A 2006 documentary called *Who Killed The Electric Car?* explored General Motors' ill-fated production of an electric car, the EV1. It was built in response to California's 1990 mandate for a zero-emissions vehicle standard; the car company eventually decided to discontinue production of the EV1 and attempted to recall all models that had been sold.

The film suggested that GM had never really tried to market the car and asked why the company had been so persistent in trying to recall the cars. GM's answer was that, despite $1bn spent on developing and marketing the EV1, only 800 people actually wanted to lease it. Moreover because demand was so low, parts suppliers began to discontinue production, which would have made it hard to continue to guarantee and repair those EV1s remaining on the road after the end of production. However, the saga has now come full circle, with a film sequel planned, named *The Revenge of the Electric Car*, now that GM is planning to mass-produce the Volt electric vehicle from 2010.

▶ **Perpetual motion** Stanley Meyer was a US inventor who claimed to be able to power a car indefinitely on a finite amount of water. His idea appeared to be a fuel cell that would break water up into hydrogen and oxygen. The hydrogen would provide the energy to power the car, and to reconstitute hydrogen and oxygen molecules back into water – thus starting the whole process again. Meyer died suddenly after at the age of 58, after a restaurant meal that led some to suggest he had been poisoned to suppress his invention. His fuel cell had, however, already been discredited in a court case brought against Meyer by one of his investors, in which expert witnesses claimed it was simply using conventional electrolysis.

A more genuinely mysterious death in the realm of energy was that of Rudolf Diesel, the German inventor of the generic engine that bears his name. He fell overboard in 1913 on a ferry from Antwerp to Harwich. Diesel had used biofuel in his first engines in the late 1880s, and it has been suggested by conspiracy theorists that he might have been murdered in order to squash biofuel's chances as a rival to oil-based diesel – the theory would have more plausibility had he actually persisted with biofuel. Other suggestions of suicide, or assassination by the German secret-service (lest Diesel tell the British about his technology) are equally unconvincing.

▶ **"Perpetual" light bulb** The light bulb that doesn't wear out is something that many people intuitively imagine ought to be easily achievable, leading to conspiracy theories that it must have been invented, only to be suppressed by the makers of bulbs which tend to go pop after a mere 1000 hours. Certainly, there have been restrictive practices in the light bulb industry. Between World War I and World War II, lightbulb manufacture was run as one big anti-competitive cartel that fixed prices, set output quotas and divided markets in a way that almost certainly also stifled innovation.

But regulators are now forcing the pace. In the European Union, for instance, the phasing out of the old-style incandescent light bulbs invented by Thomas Edison, which produce a lot of heat as well as light, began in 2009 and will be complete by 2012. We are moving, or being moved, towards compact fluorescent light bulbs. They are more efficient: a 20-watt fluorescent bulb apparently produces the same amount of lumens, the scientific measurement of light, as a 100-watt incandescent bulb (even though they often look a little dimmer). The new bulbs take more money and energy to make than the old ones, but they use less energy and last between 5 and 16 times longer. They don't last forever, but they are certainly longer-life.

▶ **Cold fusion or con-fusion?** We have already seen in chapter five how enormously difficult it is to replicate the huge power, temperatures and pressures of the sun to achieve nuclear fusion. So there is a natural temptation to welcome any easier alternative method of nuclear fusion, such as that which two University of Utah scientists appeared to offer in 1989. Martin Fleischmann and Stanley Pons created a sensation when they announced they had achieved nuclear fusion at room temperature. In a table-top experiment they reportedly carried out electrolysis of heavy water (richer in deuterium than normal water) with a palladium electrode and produced excess heat on a scale that, they said, only a nuclear reaction could explain.

Other scientists, however, failed to replicate their "cold fusion" experiment and the pair lost credibility – although not to everyone. Eugene Mallove, an experienced science writer, took up their cause. He wrote a book, called *Fire from Ice: Searching for the Truth behind the Cold Fusion Furore*. Claiming that Fleischmann and Pons had made a breakthrough in producing "over-unity energy" – getting more than they put in – Mallove argued that the pair's achievement had been disparaged by mainstream physicists trying to protect their own research and funding. Mallove died in 2004 during an apparent burglary on a property belonging to his parents, a tragedy that has encouraged the conspiracy theorist.

The new energy morality

Doing the right thing

Moral questions don't usually come up in books about energy. But energy use is largely responsible for man-made emissions contributing to climate change. And climate change can't be tackled without addressing some moral questions about energy.

They are moral questions because they involve notions of obligation – in shouldering energy and emissions costs now to prevent future generations paying a far bigger bill – and of fairness – in the just distribution of these energy and emissions costs between rich and poor in today's unequal world.

Cost-cutting across the globe

Our debt to future generations

Curbing our use of fossil fuels and developing alternative renewable energies to prevent climate change spiralling out of control squares with the ethical basis of sustainable development. This is that we should pay due regard to the welfare of future generations. In the classic definition offered by the Brundlandt Commission in its 1987 report to the United Nations, sustainable development "meets the needs of the present without compromising the ability of future generations to meet their own needs".

Given the current pace at which we are filling the atmosphere with carbon to "meet the needs of the present", one can hardly say that "the ability of future generations to meet their own needs" is assured.

Views vary on the degree of our obligation to the future. Some recognize no obligation to future generations, and in a literal sense there can be none. Such attitudes effectively ask "what have future generations done for me lately?" Nothing, of course. Yet generations unborn have a stake in our decisions, and we have a moral obligation to safeguard that stake. The Stern review on the economics of climate change implied that this obligation had no upper limit when it suggested that "we treat the the welfare of future generations on a par with our own". This stance implies that we should imagine all future generations in a room with us, far outnumbering the six billion of us, as we take today's climate action decisions. There is nothing whatsoever wrong with us maintaining a keen regard to future generations. This book is firmly on the Stern review side of the spectrum of views, but cautions that a sense of proportion is wise if our obligation to the future is to ever win wide acceptance. The pressing issue is how to distribute the costs and sacrifices involved in curbing carbon emissions and in developing low-carbon energies.

Should the rich cut their emissions to let the poor increase theirs?

On the demand side, the answer to the question above is yes. There are only so many GHGs that we can safely put into the atmosphere, and the worst-case scenario is that we have already reached that point. The industrialized countries, which created the global warming problem with their industrial revolution, need to give China, India and the developing world some headroom for further development (which will almost inevitably involve fossil fuels for the time being) towards Western levels. In order to provide this headroom, the rich world needs to cut back on its own fossil-fuel use.

This is therefore a unique point in the planet's history. Never before have all the countries of the world been effectively required to share out a set limit of a finite resource (earth's carbon absorption). Never before have richer industrialized countries had to consider accepting a levelling-down towards poorer developing countries. Granted, richer countries currently provide financial development aid to poorer countries, but they do so in the context of their own growing economies and expanding national budgets. Climate change makes things radically different. Climate stability is a zero-sum game – and cutting overall emissions to get down to a stable level will be a *negative*-sum game. In other words, any increase in emissions (by poor countries) must be matched by an even bigger decrease (by rich countries).

This all sounds very revolutionary. Yet much of it is enshrined in the 1992 United Nations Framework Convention on Climate Change, signed and ratified by 154 nations. Right at the start, this document recognized that "the largest share of historical and current global emissions of greenhouse gases has originated in developed countries, that per capita emissions in developing countries are still relatively low and that the share of global emissions originating in developing countries will grow to meet their social and development needs". The Convention laid out the basis of an eventual agreement between industrialized and developing countries "in accordance with their common but differentiated responsibilities", a phrase commonly shortened just to the initials CBDR by climate negotiators.

There are moral and practical arguments for differentiated cuts. The main moral argument is that industrialized countries bear the historic responsibility for the industrial revolution and build-up of greenhouse gases, so they should bear the brunt of cleaning up the mess. The main practical argument is that the richer and stronger economies of industrial-ized countries are better able to bear this brunt. And moral and practical factors come together in the "polluter pays" principle: that polluters ought to pay, and that having to pay is a practical incentive not to repeat the offence.

The differentiation is reflected in the 1997 Kyoto Protocol, implement-ing the UNFCCC. The Kyoto treaty requires only the industrialized coun-tries to reduce emissions, although they can meet part of their commit-ment by investing in emission reductions in developing countries under the Clean Development Mechanism (CDM). These investments earn a credit, which investors can use to meet their own Kyoto commitments or sell them on to Europe's Emission Trading Scheme. Some critics complain that the CDM offers industrialized countries a partial cop-out from the task of reducing emissions at home, but it also provides special – or dif-ferentiated – help to developing countries with emissions.

Contract and converge?

If the process of differentiated emission or fossil-fuel-based energy cuts were to continue, it could theoretically one day lead to everyone in the world having the same level of greenhouse-gas emissions per head. This is the hope of the Global Commons Institute, which came up in 1995 with their "contraction and convergence" proposal. Their idea is that overall emissions should contract to a safe level, and that per capita emissions should converge to the same level for all. It can hardly be faulted on moral grounds.

But the political feasibility of persuading north Americans, Europeans and Australasians to agree to massive cuts in emissions which, if low-carbon energy cannot match the potency of today's fossil fuels, will compromise their current lifestyles, so that China and India can raise their standard of living, is quite another matter. That is why we need action on the supply side.

A low-carbon diet for the wealthy

Low-carbon technology is the only "get out of jail" card the rich countries can play so that they can avoid having to cut their own energy use in parallel with their emissions. But the industrialized countries will also have to be far more generous in sharing low-carbon energy techniques to developing countries than they have historically been with other technologies. The large pharmaceutical companies' discounts in the price of drugs to help developing countries in order to combat AIDS and diseases such as malaria might be seen as a precedent for low-carbon technology transfer to developing countries.

Differentiation EU-style

An agreement within the 27-nation European Union in late 2008 provides one example of how differentiation can be successfully negotiated. The poorer EU members, mostly located in Central and Eastern Europe, have been given less demanding national targets, in terms of increasing renewable energy and reducing emissions, than their richer EU brethren in Western Europe. So at one extreme there is Romania, which is only required to make a 6.2 percentage-point renewable increase in its energy mix. And at the other extreme lies the UK, which has been given the biggest target – a 13.7 percentage-point increase. The same sort of differentiation exists in emissions, with poorer Eastern countries being allowed to increase emissions in sectors like transport, service and agriculture, and richer Western member states being required to cut emissions in these areas.

This differentiation among the member states was politically necessary in order to get their agreement to the collective EU goal of an average twenty percent cut in emissions and an average twenty percent renewable share of total energy by 2020.

But differentiation was also a moral imperative. Renewable energy simply costs more money than fossil fuels, and equal renewable targets for all 27 EU states would have placed a disproportionate financial burden on poorer states. (In fact, if you wanted to go about apportioning renewable energy quotas and targets in Europe in a purely cost-efficient manner, ignoring national affluence and culpability, you would give higher targets to the poorer Eastern states, because their forests, farmland and rivers contain more potential for renewable biomass and hydropower than remains in western Europe. If you went about climate change in the same way – disregarding historic responsibility for nineteenth and twentieth century industrial pollution or income differentials between countries – you would ask China to cut emissions as much as the US, and more than Europe or Japan. For China has far more potential to reduce its emissions than Europe or Japan.)

The EU is a little different from the rest of the world. In a permanent union of neighbouring countries, which supposedly aspire to common values and which have common policies to promote economic cohesion, you would expect richer states to take on more of the common burden than poorer members. Some of the EU's poorer Eastern states still grumble that they are being asked to make too rapid an energy transformation of their economies. Nonetheless, the EU has an eighth of all the world's countries in its membership, and a spread in wealth per head between richest (Luxembourg) and poorest (Bulgaria) that is wider than the income gap between the US and China. Its acceptance of common but differentiated responsibilities is encouraging.

The new energy world

A slow dawn

The sun is rising on a new energy world. For all the evidence that supports the naysayers, it surely cannot be that attempts to escape the fossil-fuel age are futile. Yet, with all of today's endless discussion of energy crises and crunches, climate policies, emission trading schemes, carbon budgets, renewable resources, new low-carbon investment frameworks, and invocations of "green" in every conceivable response to the greenhouse gas problem, it is proving to be a very slow dawn.

This is hardly surprising, given the nature of the energy industry – one whose investments are big, long-term, and constrained by both public interest considerations and fundamental laws of physics. "Scale and time are important in energy," complains Robin West, head of the PFC Energy firm in Washington DC. "They, the environmentalists, make the transition to a low-carbon economy sound so simple, but it isn't". In the absence of a major energy crisis such as, say, a revolution in Saudi Arabia and total shut-down of all its oil exports, inertia and passive resistance to change will continue to persist among a number of influential groups.

Industries

Low-carbon energy has to find its place in an economy based on fossil fuels. A few years ago there was the occasional sign that businesses centred upon fossil-fuel exploitation might take the initiative in wholeheartedly transforming themselves into low-carbon energy businesses.

In 1997, under CEO John Browne, BP broke ranks with the rest of the oil industry. It came out in favour of taking action against climate change, started an internal system of carbon trading, began to invest in solar and wind, and re-branded itself as Beyond Petroleum to symbolize its focus on cleaner energy.

Some other oil companies followed BP some of the way in this direction. But by 2009, Browne's successor Tony Hayward had reorientated BP's focus back on fossil fuels. He reacted to a temporary global glut in solar components by cutting back BP's solar business, and responded to wind farm planning delays in the UK by focusing BP's wind investments in the US. More ominously, his stewardship has seen BP beginning to explore Canadian tar sands – a high-carbon energy source Browne had avoided. BP has retained the slogan Beyond Petroleum, but this only tends to invite criticism of hypocritical "greenwash".

Perhaps it was always too much to expect any oil company to resist the lure of the upstream – and the high profits from extracting oil and gas there – that will always interest their shareholders most. But it is a pity that the oil companies are generally beating a profitable retreat back to their "core" hydrocarbons business, and failing to put their project management skills and offshore experience to good use in setting up big wind farms at sea, for instance.

Of course, the oil companies are far from the entirety of the energy industry, which has thousands of general utility companies providing electricity and gas by many different means. They are beginning to constitute an enormous vested interest in low-carbon energy. HSBC Bank runs a Climate Change Index on which it lists businesses selling low-carbon goods and services. In 2009 HSBC reported that the "clean technology" sales of these listed companies had risen to $534bn in 2008, more than the $530bn global turnover of the combined aerospace and defence sectors in that year.

But the low-carbon parts of the energy industry are frequently at odds with each other. In many countries the renewable and nuclear lobbies are each warning governments about the other. Renewable energy producers say nuclear will suck up all the public money, and leave none for them. Nuclear energy companies point out that renewable energy development will lower the carbon penalty that is their only advantage over fossil fuels.

Individuals

No previous energy revolution has ever put more financial, and moral, burden on consumers' shoulders as the low-carbon transformation would – for the simple reason that no energy revolution has ever put so much emphasis on consumers using energy more efficiently. Efficiency is upfront money – *our* upfront money. We are increasingly bombarded with messages to buy hybrid cars, air-source heat pumps, condensing boilers and energy-saving fluorescent light bulbs. These currently all cost more to buy than the item they replace. Yes, there is a pay-back over the lifespan of these products because they use less energy. But it is likely to take several years. So while people still have a choice (with the exception of light bulbs, as old-style incandescents come off the market) they will dither about hybrid cars and condensing boilers.

Granted, governments go about their ponderous business of proposing new efficiency standards, putting them out to consultation with the public and industry, and eventually legislating. But there is no outside force to impart any urgency to energy efficiency. Greenpeace, for instance, is an environmental NGO whose members' stock-in-trade is direct action. But short of breaking into people's houses and forcibly insulating wall-cavities and lofts, there is not much direct action they could take.

Governments

So what are governments doing to counter this inertia and accelerate the pace of low-carbon energy change? Their one big thing has been the green part of the fiscal stimulus to counter the 2009 recession. HSBC Bank estimates that the climate-change-related part of the stimulus (including water and waste treatment, as well as low-carbon energy and energy efficiency) totals $512bn. That's hardly peanuts. Although by November 2009 only a little over ten percent of this promised cash injection had actually been spent (and most of that by China), the green stimulus may help the major economies come out of recession better adapted to, and skilled in, low-carbon energy than when they crashed into it. That is the aim. Every recession alters a country's economic structure, and the goal of the green stimulus is to steer it in a green direction.

But we should have no illusion that it was only the shock of impending financial collapse in autumn 2008 that set governments on this course. Governments needed that shock to act. And the shock is wearing off. The question now is what will jolt, and keep jolting, governments into sustained low carbon action.

Here we should be careful what we wish for. The veteran US environmentalist Lester Brown has discussed different scenarios that might mobilize climate action in terms of what has jolted the US into action in the past. He suggests that ecological equivalent of the Japanese attack on the American navy in Pearl Harbor, prompting the US to enter World War II, might be a major break-up of the west Antarctic ice sheet which "could raise sea level a frightening two or three feet with a matter of years". Unfortunately, as Brown points out, "if we wait for a catastrophic event to change our behaviour, it might be too late".

Brown has two other scenarios for social change. One is "the Berlin Wall model", wherein "a society [in this case, eastern Europe] reaches a tipping point on a particular issue – often after an extended period of gradual change in thinking and attitudes." The other is the "sandwich model", where "there is a strong grassroots movement pushing for change on a particular issue that is fully supported by strong political leadership at the top". Brown's example of the Berlin Wall model working in the US is the shift of opinion against smoking tobacco. This shift occurred over several decades, reaching a tipping point in the last 10–15 years, after which controls on sale and advertising tobacco soon followed. However, the parallel understates the importance of climate change. While smoking grows more dangerous over time to the individual smoker and raises overall health costs, it does not get dramatically more dangerous to society as a whole over time, as climate change does.

So Brown argues that the sandwich model of strong grassroots pressure – hopefully working with, rather than against, political leadership – is more appropriate to the more urgent problem of energy-induced climate change. The expedient tactic of government ministers inviting environmentalists to pile pressure on their own governments has worked in the past, as shown by the success Friends of the Earth's campaigning in the UK. At present, most people in the industrialized world are more preoccupied with recession. Nonetheless, if there were greater mobilization on climate change, those mobilized would very likely find themselves leaning on ministers' open doors.

So what should individuals push for?

Here are a few suggested principles:

Lead by individual example

The point of taking individual action on energy is to make a difference, where possible, in saving, efficiency and perhaps even generation – but even more to build the idea that energy waste is socially unacceptable (as has happened with tobacco). In his otherwise brilliant book, *Sustainable Energy – Without the Hot Air*, David MacKay disparages the notion in "every little helps" in energy saving and efficiency. He argues that "if everyone does a little, we'll only achieve a little".

But while this is a useful warning against complacency, and a call to scale up low-carbon power, it ignores the possibility of individual action contributing to the spread of societal norms about energy saving.

Collective action

There are some excellent examples of collective action making a difference that is larger than the sum of its parts. One is of wind-farm cooperatives in which local people and firms have banded together to achieve greater scale and power than they could individually. The wind power generated by a turbine is proportional to the square of the turbine's diameter (see p.119). Another pertinent example is the Carbon Trust's minor coup in persuading members of the British Frozen Food Federation that they could all reduce their energy consumption by turning down the temperature in their supermarket freezers a couple of degrees without the produce going off.

Paying the full cost for energy

The EU nations committed to cutting their emissions by 50 percent by 2050 (or 80 percent in the UK case). It is hard to imagine how this can be achieved without the public paying more for their energy. At the same time, as democracies supposedly committed to social welfare, they have a responsibility to help those who can least afford more expensive energy bills. In this regard, the government payment schemes for older cars to be scrapped – called "cash for clunkers" in the US – have proved a winner.

Of course, these schemes were introduced in 2008–09 mainly to help various car industries around the world. But older cars tend to be driven by poorer people who, under these schemes, received money towards newer and cleaner models they could not otherwise afford.

Telling the government to intervene

Governments have long been active in regulating energy-hungry products, but they have only recently woken up, or returned, to the realization that the market isn't delivering the low-carbon economy as fast as they need it to. After the discrediting of financial markets during the recession, taboos are disappearing, even in the UK and US, against non-market mechanisms like targets and quotas for renewable energy or taxes and minimum prices to prop up carbon prices. Most governments have never been more susceptible, even amenable, to pressure from their public.

Don't let the best be the enemy of the good

While many climate-change activists might like to see all fossil-fuel and nuclear projects blocked forthwith, it would be sensible to rely more on nuclear power and perhaps gas, the cleanest of fossil fuels, in order to speed up reduction or elimination of coal, the dirtiest hydrocarbon.

Only the combined efforts of the nuclear and renewable energy sectors are likely to get us to a low-carbon economy. Natural gas could, likewise, be an essential temporary bridge to that low-carbon nirvana. Pragmatism, rather than dogmatism is necessary.

Don't be exclusive

Remember, too, that a properly workable energy policy for the future will be composed of a multiplicity of energy sources and efficiencies. It will be a policy of this … and this … and this … and this … Hopefully, this book will have helped to fill in some of those dots.

RESOURCES

Books

Hubbert's Peak: the Impending Oil Shortage
Kenneth Deffeyes (2001)
Deffeyes, a one-time colleague of Hubbert's at Shell, takes the techniques of M.King Hubbert in correctly forecasting the peak in US oil, and applies them to world oil. His predictions are off target (so far), but his exploration of the issues is spot on.

Ten Technologies to Save the Planet
Chris Goodall (2008)
This covers the most important renewable technologies. It has the rare merit of explaining with clarity how these technologies work and with realism the uphill task of displacing fossil fuels.

Blackout
Richard Heinberg (2009)
The latest work by this prolific American journalist and educator focuses on coal as the biggest menace to the climate, and explodes the myth of coal as provider of long term energy security.

The New Energy Paradigm
ed. Dieter Helm (2007)
An overview of the current energy policy debate, addressing environmental issues and considering how international energy markets effect security of supply.

The Rough Guide to Climate Change
Robert Henson (2006)
Everything you need to know about climate change.

The Transition Handbook: from oil dependency to local resilience
Rob Hopkins (2008)
This a green living manual, designed to improve local communities' ability to withstand the shock of the peak oil to the global economy.

Sustainable Fossil Fuels: the Unusual Suspect in the Quest for Clean and Enduring Energy
Mark Jaccard (2005)
The case for the potential of fossil fuels to be turned into clean, sustainable energy sources.

Energy and Security: Toward a New Foreign Policy Strategy
ed. Jan H. Kalicki (2005)
A collection of essays by different experts examining the relationship of the US with the producer nations that it depends upon.

Blood and Oil
Michael Klare (2004)
This explores the dangers of America's growing petroleum dependency, which has economic and foreign policy consequences for us all.

RESOURCES

The Carbon War
Jeremy Leggett (2000)
This is a racy account of the politics of global warming and the Kyoto protocol negotiations by someone who has been successively academic, oil geologist, Greenpeace director, and now solar power entrepreneur.

Sustainable Energy – without the hot air
David MacKay (2009)
This Cambridge physicist's mission is to bring numeracy to the debate, so as to help people judge whether the claims and counter-claims about renewable energy add up or not. It is a brilliant book, but you don't have to be brilliant to understand (most of) it.

The Final Energy Crisis
ed. Andrew McKillip with Sheila Newman (2005)
An international collection of essays examining various aspects of over-reliance on fossil fuels.

Future Energy: How the New Oil Industry Will Change People, Politics, and Portfolios
Bill Paul (2007)
If energy demand, as many suggest, is one of the best business opportunities of all time, what will people be investing in?

The Grid
Philip Schewe (2007)
A readable account of the extraordinary spread of electricity. It is US-focussed, but relevant to all countries with modern grids.

Energy
Vaclav Smil (2006)
This is one of many energy-related books by Smil, a Canadian author who is particularly good on the science of energy.

The Economics of Climate Change
Nicholas Stern et al (2006)
This report, commissioned by the UK government, was the first big economic analysis of climate change. Its political impact lay in showing that the costs of early climate action are vastly outweighed by the benefits.

The Prize: the Epic Quest for Oil, Money and Power
Daniel Yergin (1991)
This is still the most readable general account of the oil industry's first 130 years.

Websites

General information

How Stuff Works www.howstuffworks.com

As you might expect, this provides detailed explanations of how things work, from the electric grid and photovoltaic cells to fuel cells and wind turbines.

Governments and academics

The Energy Information Administration www.eia.doe.gov
The information and statistical arm of the US Energy Department. It also has some useful explanations of aspects of the energy industry and energy markets.

The European Commission (and other EU institutions) ec.europa.eu/energy
Euractiv www.euractiv.com/en
The UK Department of Energy and Climate Change www.decc.gov.uk
The three websites above are invaluable for keeping up with EU energy, climate and other related policies.

Intergovernmental Panel on Climate Change www.ipcc.ch
The IPCC is a UN body whose role is to assess the scientific basis of human-induced climate change by assessing scientific, technical, and socio-economic research.

The International Energy Agency www.iea.org
The main energy organisation of the industrialized, mostly oil-importing countries. It publishes the comprehensive annual forecasting book, *World Energy Outlook*.

The Tyndall Centre for Climate Change Research www.tyndall.ac.uk
Academic centre promoting a sustainable response to climate change in the UK and internationally.

United Nations Development Programme www.undp.org/energy
Home of the United Nations' energy-related sustainable development projects.

US Department of Energy www.energy.gov
A prime source for statistics and reports on energy research commissioned by the US government.

The energy industry

BP Statistical Review www.bp.com
Compiles data voluntarily submitted from right across the energy industries.

European Nuclear Society www.euronuclear.org
International federation of 26 societies that promote and advance science and engineering for the peaceful use of nuclear energy.

Hubbert Peak of Oil Production www.hubbertpeak.com
Data, analysis, and recommendations regarding the "impending peak of global oil extraction".

The Organisation of Petroleum Exporting Countries www.opec.org
Opec's website has some basic information on oil producing countries.

World Business Council for Sustainable Development (WBCSD) www.wbcsd.org
A CEO-led, global association of around 200 companies dealing with business and

sustainable development. The Council provides a platform for companies to explore sustainable development. Publishes an annual review and other statistics.

Campaigners, pressure groups and NGOs

Climate Action Network www.climatenetwork.org
A worldwide network of over 365 Non-Governmental Organisations that promote action to keep human-induced climate change within ecologically sustainable levels.

Friends of the Earth www.foei.org

Greenpeace www.greenpeace.org

Blogs

The many interesting and useful UK bloggers on energy-related issues include:

Jeremy Leggett www.jeremyleggett.net

George Monbiot www.monbiot.com

Stephen Tindale climateanswers.info

Excellent links to US-focussd energy bloggers can be found through the following pair of websites:

Energy Authority www.energyauthority.net; The Post-Carbon Institute www.postcarbon.org

DVDs

A Crude Awakening
Frank Messmer's documentary examining peak oil theory and the impact it would have on our oil-dependent society.

The 11th Hour
Narrated by Leonardo DiCaprio, this doc picks up where *An Inconvenient Truth* left off by presenting solutions to climate change.

The End of Suburbia: Oil Depletion and the Collapse of the American Dream
An explanation of how the American lifestyle has become dependent on cars and oil.

An Inconvenient Truth
Al Gore's landmark documentary explaining what climate change is and how we can stop it.

Who Killed the Electric Car?
Chris Paine and Martin Sheen's documentary looks at how the electric car was developed and subsequently dropped in the late 1990s, exploring the economic, scientific and political reasons.

INDEX

T

Picture Credits

Picture Credits